Air Engines

Air Engines

The History, Science, and Reality of the Perfect Engine

Theodor Finkelstein

and

Allan J Organ

ASME PRESS

The American Society of Mechanical Engineers, New York

Published with corrections in 2009 by ASME Press.

First Published 2001 in the United Kingdom by Professional Engineering Publishing Limited and in the United States by ASME Press.

ISBN-13: 978-07918-0171-0

ISBN-10: 0-7918-0171-3

ASME Order No. 801713

A CIP Catalogue record for this book is available from the British Library.

Cover illustration

Those familiar with the first printing of *Air Engines* will recall its distinctive jacket illustration. The 6-cylinder concept under development by Dipl.-Ing Peter Feulner, together with accompanying background information, was a valuable complement to the material between the covers. The original art-work was not preserved, however, and to the time-scale of this re-print it has not been possible to arrange for re-use of the illustration.

The replacement jacket design makes an important symbolic statement: it is a graphic generalization of the conversion algebra defining the equivalence between thermodynamic volume ratio κ and kinematic displacement ratio λ, and between thermodynamic phase angle α and corresponding kinematic phase angle β. Equally it defines an essential condition for a coaxial ('beta') engine to be thermodynamically equivalent to a given 2-cylinder ('alpha') engine - and *vice-versa*. (The text has more on this radical unifying insight at Chap. 6.)

Ted Finkelstein casually knocked off the crucial algebra in the course of his *tour de force* opus '*Optimization of phase angle and volume ratio for Stirling engines*'. With the companion paper '*Generalized thermodynamic analysis of Stirling Engines*', this material inaugurated the modern analytical base of Stirling engine technology. To this extent, the jacket illustration symbolizes a defining step in the progression from the air engine of the Industrial Revolution to the Stirling engine of today - and from chapters 1 - 4 of the present account to the chapters which follow.

The reader encountering *Air Engines* for the first time is encouraged to become apprised of the Feulner multi-cylinder concept. A computer-generated solid-model image and an informative link can be found on the *Welcome* page of *http://web.me.com/allan.j.o/Communicable_Insight*

The scotch yoke and the eccentric mechanism invented by Sir Charles Algernon Parsons have been identified as viable solutions. The latter is based on the hypocycloid that degenerates into a straight line for a radius ratio of 1:2. Both scotch yoke and Parsons' mechanisms produce simple harmonic motion, thereby making possible perfect balancing through use of simple bobweights. (The rocking couples due to the pairs of cylinders being situated in different planes remain unbalanced.)

The rigid coupling between two opposing pistons gives freedom of choice of piston rod diameter, thereby enabling considerations of thermodynamic swept volume ratio to prevail over those of bearing loads etc.

Thermodynamic phase angle leads to two overall layouts: the four-cylinder (thermodynamic phase angle, $\alpha = 90°$) and the six-cylinder ($\alpha = 120°$).

The evolving engine concept lends itself to a consolidated heater design for all gas circuits (single row tubular) which, for viable power ratings, leads to a combustion chamber with the aspect ratio that has shown characteristics favouring low NO_x formation.

Cycle simulation studies based on V Eppert´s work at Kassel University suggest that the engine concept is appropriate to power ratings from 15 kW to 100 kW. A power to weight ratio of 1 kW/kg can be expected above 50 kW rated output.

The radial six has been conceived along with a 'narrow V', five-cylinder version of equal power. Neither design is yet fully engineered, but it is worth noting that they draw on comprehensive experience of an earlier engine designed by Peter which was built, comprehensively instrumented, and extensively tested at Kassel University. A massive accumulation of experimental data remains to be interpreted and published. Interested readers will find further reference to the Kassel programme in student dissertations from the University.

Peter holds the post of engineering project leader with Ingenieurgesellschaft für Fahrzeugtechnik mbH, Leinzell, Germany, for whom he carries out design and development consultancy for the automotive industry. Over and above his full-time profession and his involvement in Stirling engines he is a serious student of philosophy with a particular interest in the approach of A N Whitehead – possibly explaining a command of English surpassing that of many native speakers. A contact e-mail address is feulner@iftmbH.de

Related titles

For a full range of titles available from ASME International, please visit the ASME Press website at http://www.asme.org/Publications/Books/

About the authors

Theodor Finkelstein studied mechanical engineering at Imperial College, London, obtaining his BSc with first-class honours. A Diploma of Imperial College followed. Working under Professor H Heywood he was awarded his PhD in December 1952 for a thesis entitled *Theory of Air Cycles with Special Reference to the Stirling Cycle*. He is a registered Professional Engineer.

He has worked mainly in the research departments of major corporations, but has also held academic appointments at the Universities of Wisconsin, San Fernando Valley State College, University of Calgary, and University of California.

He pioneered the development of mathematical models of the Stirling cycle and the use of electronic computers, both analogue and digital, for obtaining solutions. His analytical optimization of the Schmidt cycle model has never been superseded.

Inventions, some patented, have included the internal heat supply and transmission (1958) in the free piston machine (1960), balanced compounding (1975), and the introduction of the Delta class of Stirling engines (1998). He has contributed an average of one scientific paper per year over a period of 50 years to the Stirling literature. In 1978, he and Stig Carlqvist founded Stirling Associates International which continues to carry out consulting work in this field.

His later activities have concentrated exclusively on industrial Stirling engine development. He lives with his wife, Hanna, in Beverly Hills, California.

Contact details:
PO Box 643, Beverly Hills, California, 90213, USA
Phone: (USA) 310 472-2176
Fax: (USA) 310 476-2021
e-mail: StirlingResearch@aol.com

Allan J Organ studied for his BSc in the Department of Mechanical Engineering at Birmingham University, where a final year project provided an irresistible introduction to the Stirling engine. Study of the regenerator began at Toronto under Professor F C Hooper and led to the degree of MASc in 1964. He returned to Birmingham to take up a research position studying metal-forming at high rates of strain under Professor S A Tobias, for which he gained a PhD in 1968. Metal-forming provided the bread and butter until 1970, when a British Council Visiting Professorship in Brazil gave the opportunity to divide the research interest between plasticity and Stirling engines.

The return from Brazil was to a position at King's College London allowing free rein of research activity, and it is not necessary to say which way the choice went! A privilege of a further move to Cambridge in 1978 was also complete freedom of research interest. The die being well-and-truly cast, attention turned to making Stirling engine analysis less restricted to a specific engine and more relevant to the genre as a whole. A first book, published by Cambridge University Press in 1992, approached this task from the standpoint of Dynamic Similarity, but had gone to press before the full benefits of this powerful tool had been completely perceived and applied to best advantage. Attention reverted to the regenerator problem, which finally yielded in 1995. This afforded an opportunity for drawing together gas circuit design and regenerator theory using the full power of Energetic Similarity. The text was published in 1997 by Mechanical Engineering Publications (MEP).

He was awarded the post-doctoral degrees of DEng from the University of Birmingham in 1993, and ScD by Cambridge in 2000. In addition to the two books, he has contributed 41 scientific and 10 technical papers to the literature on Stirling engines and the regenerator, and 15 papers to metal-forming literature.

He is a Fellow of the Institution of Mechanical Engineers and serves on the Editorial Board of the *Journal of Mechanical Engineering Science* (Proceedings Part C). He lives in the village of Dry Drayton near Cambridge.

Contact details:

32 Pettitt's Lane, Dry Drayton, Cambridge CB3 8BT
Phone: (UK) (0)1954 781815
Fax: (UK) (0)1954 781253
e-mail: allan_j_organ@talk21.com

Contents

This book is dedicated to the genius of Robert Stirling

Foreword

The title *Air Engines* is that of the classic series of articles that form the first four chapters of the book. But it is evident on reading that there are other reasons for preferring this over the more obvious *Hot-air Engines*. The most important is the dramatically improved performance of the modern, air-charged, external combustion engine relative to that of the hot-air toy or the 19th century industrial engine. Indeed, the air engine promises to challenge the specific power and efficiency of hydrogen- and helium-charged types.

Hot air (of the metaphorical variety) is an unfortunate feature of much of the general literature on this particular technology, which carries a disproportionate burden of ideas that have never been tested, or of which test results have – for some reason – been withheld. Be assured that this book contains no such hot-air. The fact is guaranteed by the well-known names of the authors, Allan J Organ and Theodor Finkelstein, who together can account for nigh-on a century of research and design work in this field. I spent two summers at Cambridge as a student of Dr Organ before being truly satisfied with my own insight. Perhaps I can offer the excuse that this book was not available at that time.

The treatment opens with a classification of air engines as a basis for an account of their conception and historical evolution. Inevitably, attention soon focusses on Robert Stirling's remarkable double invention – the thermal regenerator and the famous engine of 1816 – as being the only viable basis for a modern prime mover. The story of the air engine is essentially that of the development of modern thermodynamics from its origins in Caloric theory. Finkelstein masterfully combines the two aspects so that, when Chapter 4 concludes with the re-birth of the modern Stirling engine at the hands of Philips, you have a comprehensive insight of the thermodynamic and mechanical design aspects without use of a single mathematical symbol!

Structurally the book may be unique in consisting of two sections, the early chapters written 40 years before the rest. It is a tribute to his handling of the material that Dr Finkelstein's contribution is as pertinent now as when written, and appears in essentially unmodified form. Only a brief regression is called for in the bridging chapter to draw attention to early developments that have come to light since 1959. Subsequent chapters bring the account up-to-the-minute, continuing in the process to highlight the relationship between technical progress and the evolution of theoretical understanding.

Superimposed on the chronological account is an explanation of thermodynamic design tools which can be applied to your own engine. There are details of air engines currently under development for applications especially suited to their particular characteristics.

The book offers an easy-to-read, coherent, and scientific account based on the authors' pioneering development of (dimensionless) *scaling parameters*. This amounts to the distillation of practical and theoretical know how and experience that has previously appeared over many years in high-level papers and books and which is confirmed by the latest research results.

The slow progress of the air engine towards widespread commercial exploitation is due, more than anything else, to a lack of *communicable insight*. An evident priority of the book is to make good this lack. Thus it provides a long-overdue basis for communication among air engine experts and hobbyists, serving for practical design or as a reference text in research and academic work. A 'forensic' enquiry into the original engine of 1818 opens up a new dimension in the technical history of air engines.

I am deeply impressed by the work of the authors. But my sympathy is also with a vision urgently required by my generation, and upon which this topic bears: there is a pressing requirement for an environmental friendly, renewable global energy conversion system having limited risk. My current view is that the air engine has a key role in CHP (Combined Heat and Power) systems, including solar dish and biomass-fuelled types. The technology is now in place, but successful implementation awaits the formation of a critical mass of funding, manpower and, above all, of *communicable* insight. This book takes a step in this direction.

If you think the air engine just functions by moving a gas between a hot and a cold space you will find there is more to explore than you ever expected before opening this book. Whatever your previous technical background you will watch an air engine with greater fascination the next time around – even if it is, indeed, a toy of the 'hot-air' variety.

Dipl.-Ing. Peter Mäckel
Kassel
peter@maeckel.com

Preface to the original *Air Engines*

In these articles a survey of the development of air engines is given, starting with the primitive 'furnace gas engine', built at the beginning of the nineteenth century, and leading up to the modern single-stage Philips' air liquefiers. The different types of air engines are classified into three groups: open-cycle internal combustion, open-cycle external combustion, and closed-cycle engines, invented respectively by Cayley in 1807, Ericsson in 1840, and Stirling in 1816. The principle of operation for these is explained and typical examples of each are given. Engines ranging from a small toy motor to a monstrosity with four 14 ft diameter cylinders, running at 9 r/min, are described.

To the writer's knowledge this is the first time that a comprehensive set of illustrations and descriptions of air engines has been assembled – in fact, a picture of Stirling's original engine of 1816 has previously appeared in print on only one occasion. Recent new developments are dealt with and some interesting proposals for future applications of these basic scientific principles are made.

Theodor Finkelstein
1959

Editorial Note

The aim has been to retain *Air Engines* in a form as close to the original as possible. Nevertheless, some minor departures will be noted. There has, for example, been difficulty in securing copyright permission for certain of the original diagrams. Similar figures from known sources have been substituted where available, and any explanatory notes adjusted accordingly. The note 'source unknown' against a figure indicates that no substitute could be found. (Should the reader know the origin of any of these figures the authors would welcome notification so that due acknowledgement can be included in any future edition.)

The original account did not have access to the illustration of Cayley's air engine of 1807. This has since come to light, and now complements the diagram of the 'gradual combustion' engine which earlier deputized for it. More is now known about Robert Stirling's original patent specifications (of which there was a Scottish and an English version), calling for adjustment to a few words of the narrative. A change to the Harvard system of referencing *(Author, date)* has required reallocation of those titles originally grouped under a single numerical (65) reference.

Otherwise, the text is as originally published. It is now reproduced by kind permission of the Editor of *The Engineer.*

Preface to the joint edition

Why another book on Stirling engines?

There are, no doubt, authors who can cite one clear motive. Most of us, however, find book writing a soul-wrenching experience, and need to be induced by a combination of incentives.

First, I had for decades been trying to persuade Ted Finkelstein to bring his unique series of articles in *The Engineer* of 1959 up to date, put it between covers, and thereby perhaps achieve a measure of long-overdue credit for his imaginative and independent contributions to the field. Persuasion failed, but he kindly encouraged me tackle the task on our joint behalf. Whether the result has the charisma it would have carried at the hands of the original master remains to be judged, but at least there is a result.

Second, an intriguing mystery offers itself for investigation: accounts of the modern technological development of the Stirling engine contain repeated allusion to the crucial role played by the science of thermodynamics, heat transfer, and fluid flow. Perversely, details have been consistently withheld. Even Hargreaves' otherwise definitive account* is studiously silent on this point. A potential legacy of thermodynamic design guidelines is missing, together with any accompanying insight this might have afforded into the *thermodynamic personality* of the Stirling engine. A narrative chronology of twentieth-century developments would shed little light on this unsatisfactory situation which, therefore, provides the considerably more interesting exercise of reading between the lines. This being the case, the opportunity is taken of probing a further, perplexing aspect of recent history, namely, the role of regenerator theory. The nineteenth century saw the regenerator itself – that crucial, central component – omitted or relegated to a token function. The twentieth century produced a huge corpus of analytical work on the regenerator massively outweighing Stirling cycle analysis *per se*, and continued the earlier tradition by assiduously ignoring it for engine simulation and design.

Examining these events permits reconstruction of the thermodynamic design methodology that might have been. This in turn offers the third incentive (if one were needed) for undertaking the present account: to reconstruct the thermodynamic personality of the celebrated engine of 1818, and thus to examine claims for the 2 h.p. output.

Finally, there is the matter of air engines – not hydrogen- or helium-charged engines, but air engines – their heyday past, and their heyday to come.

*Hargreaves, C. M. (1991) *The Philips Stirling Engine* Elsevier, Amsterdam.

Gathering together the various threads leads to the conclusion that the air- or nitrogen-charged engine has been wrongly set aside in the rush to exploit the apparent superiority of machines charged with hydrogen or helium. A further aspect of Stirling's original vision which has gone by the board is its elegant simplicity. Reverting to the earlier priorities, but this time with the aid of appropriate thermodynamic design tools, has shown how an air- (or, equally, nitrogen-) charged engine can out-perform the helium-based counterpart in the rapidly evolving technology of domestic Combined Heat and Power (CHP) – a role to which the operating characteristics of the Stirling engine are uniquely suited. Apparently the economics are well-matched also.

Some two centuries later, the commercial future appears to lie, after all, with an engine little changed from the original concept. Hence retention for this book of the title of the pioneering articles which form its foundation and inspiration: *Air Engines*.

Allan J Organ

Notation

(Note: The notation for the kinematics of Stirling's 1818 engine used in Table 6.1 is defined in Fig. 6.5, and is not duplicated here.)

a	Acoustic speed $\sqrt{\{\gamma RT\}}$	m/s
a, b, c, d	Dimensionless parameters of heat transfer and flow friction correlations	
a/f	air-to-fuel ratio (by mass)	
A_{ff}	Free-flow cross-sectional area	m^2
b/s	bore-to-stroke ratio	
c	specific heat	J/kg K
c_p, c_v	specific heat at constant pressure and at constant volume respectively	J/kg K
c_w	specific heat of matrix material	J/kg K
C_f	friction factor ratio of shear stress at wall to $\frac{1}{2}\rho \underline{u}^2$	
d_{FL}	readjustment interval of fluid temperature profile, defined in text	m
d_w	diameter of individual wire of regenerator gauze	m
d_{warp}, d_{weft}	diameters of wires in warp and weft directions respectively	m
D	operator – substantial derivative ($D/dt = \partial/\partial t + u\partial/\partial x$)	s^{-1}
D	internal diameter of cylinder	m
f	cyclic frequency	s^{-1}
f_{NT}	function of temperature ratio, N_T, defined in text	
f_{NTCR}	function of $¶_v$, γ, N_{TCR} defined in text	
h	coefficient of convective heat transfer	W/m^2 K
H_b	enthalpy per blow based on temperature difference ΔT	J
j	Colburn j-factor N_{st} $N_{pr}^{2/3}$	
k	thermal conductivity	W/m K
L_r	passage length in regenerator packing	m
m	variable mass	kg
m'	mass rate	kg/s
m_w	mesh number of gauze (wires/m)	m^{-1}

m_{warp}, m_{weft}	mesh number in warp or weft direction, inverse of pitch distance	m^{-1}
M	mass of working fluid taking part in cycle	kg
n_v	number of valves/cylinder of four-cycle internal combustion engine	
N_B	Beale number, $W_{brake}/p_{ref}V_{sw} = power_{brake}/p_{ref}V_{sw}f$	
N_F	Fourier modulus, $\alpha_w/\omega d_w^2$	
N_{FL}	flush ratio: ratio of mass of fluid through regenerator per blow to mass of fluid contained in matrix	
N_{MA}	characteristic Mach number, or dimensionless speed, $\omega V_{sw}^{1/3}/\sqrt{RT_{ref}}$	
N_{otto}	$power_{otto}/p_{ref}V_{sw}f$	
N_{pr}	Prandtl number $\mu c_p/k$	
N_{re}	Reynolds number, $4\rho\lvert\underline{u}\rvert r_h/\mu$	
N_{st}	Stanton number, $h/\rho\lvert\underline{u}\rvert c_p$	
N_T	characteristic temperature ratio, T_E/T_C	
N_{TCR}	thermal capacity ratio $\rho_w c_w/\rho c$ for incompressible case or $\rho_w c_w T_{ref}/p_{ref}$ for compressible case. Respective numerical values differ by the factor $(\gamma-1)/\gamma$	
N_{TF}	Finkelstein number, $W_{computed}/p_{ref}V_{sw} = power_{computed}/p_{ref}V_{sw}f$	
N_{Tr}	regenerator heat transfer scaling parameter $(L_r/r_{hr})^{3/2}N_{SG}^{-1/2}$	
N_{SG}	characteristic Stirling number or dimensionless pressure $p_{ref}/\omega\mu_{ref}$	
NTU	number of transfer units, $N_{st}L_{reg}/r_h$	
p	absolute pressure	Pa
pw, perim.	wetted perimeter	m
q'	heat rate	W
Q'	volume flowrate	m^3/s
Q_r	heat stored in regenerator per cycle	J
r	compression ratio V_{max}/V_{min}	
r_h	hydraulic radius, free-flow area/wetted perimeter	m
R	gas constant for specified gas	J/kg K
S	stroke (of piston or displacer)	m
S	entropy	J/K
t	time	s

T	absolute temperature	K
T_g, T_w	local, instantaneous absolute temperature of fluid and wall respectively	K
u	particle velocity	m/s
$\underline{u}$	mean mass velocity or bulk velocity, $m'/\rho A_{ff}$	m/s
V_d	dead or unswept volume	m^3
V_E	modulus of expansion space volume variation	m^3
V_{sw}	swept volume	m^3
W	work per cycle	J
W'_f	friction power, or fan power according to context	W
α	thermodynamic phase angle, the angle by which events in expansion space lead those in compression space	
α	coefficient of thermal diffusivity	m^2/s
α_{ff}	dimensionless free-flow area, $A_{ff}/V_{sw}^{2/3}$	
β	kinematic phase angle, angle by which motion of displacer leads that of piston in displacer-type machine	
γ	isentropic index, ratio of specific heats c_p/c_v	
δ	dimensionless dead space, V_d/V_{sw}	
ΔQ_r	regenerator heat deficit, difference between heat deposited and retrieved	J
ΔT	$T_g - T_w$	K
ε_w	wall temperature excursion as fraction of T_C	
ζ_f, ζ_{sf}	specific work per blow lost to pumping	
ζ_q, ζ_{sq}	specific work per blow lost to heat transfer in regenerator	
η_T	regenerator temperature recovery ratio, e.g. $(T_g^{exit} - T_E)/(T_C - T_E)$	
κ	thermodynamic volume ratio, V_C/V_E	
λ	kinematic displacement ratio, ratio of displacement of work piston to that of displacer in displacer-type (beta) machine	
Λ	Hausen's 'reduced length', equivalent to NTU	
μ	coefficient of dynamic viscosity	Pa s
ν	dead space parameter $\delta_r \log_e N_T/(N_T - 1)$	
Π	Hausen's 'reduced period', equivalent to $N_{FL}\ NTU¶_v/[(1-¶_v)\ N_{TCR}]$	
ρ	density	kg/m^3

τ	$T/T_{\mathrm{ref}} = T/T_{\mathrm{C}}$	
ϕ	crank angle or dimensionless time, ωt	
ω	angular frequency	rad/s
$¶_{v}$	volumetric porosity, (volume wetted by fluid)/ (envelope volume)	

Subscripts

b	value per blow
E, e, C, c	expansion, compression
f	friction or fan as per context
g, w	gas (or fluid), wall (or wire)
p, d	piston, displacer
r	regenerator
otto	relating to Otto cycle
xe, xc	expansion/compression exchanger
I, II	forward, reverse blow
$\propto$	environment or free-stream

Superscripts

+	additional (as in dead space)
'	(prime) per unit time

1

Air engines

1.1 Introduction

The 'hot-air engine', or 'caloric engine' as it was often called, has been considered obsolete until recently. It is not generally known that during the latter part of the nineteenth century many thousands of several kinds of such engines were produced and used in many workshops as the only source of mechanical power, before becoming displaced by the development of modern high-speed internal combustion engines of much greater specific power. Modern textbooks in thermodynamics or college courses on heat engines include little more than a brief reference to regenerative air cycles. Hence, very few young engineers are fully conversant with the working principle of regenerative air cycles with reciprocating mechanisms, and fewer still have had an opportunity of seeing such an engine in operation.

Extensive new developments have taken place recently, and it is possible that the old air engine is due for a revival in modernized form. It has been stated that the use of modern materials, such as light alloys and heat-resisting steels, and the application of new knowledge of thermodynamics and heat transfer, have put the whole subject in a new perspective. Claims have been made for engines with overall efficiencies and specific powers exceeding those of internal combustion engines, and a reversed air engine operating as a heat pump, in which air can be liquefied without pre-compression, is being marketed.

Many research workers in this field are beginning to realize that the inherent high thermodynamic efficiency of engines working on reversible air cycles offers possibilities which may be unattainable by any other known process. It is hoped that this technical review of past achievements in this field will stimulate further interest in the subject and perhaps prepare the ground for the application of the principle to be described to a variety of apparently unrelated purposes such as power drives, small electric generators, solar energy convertors, refrigerators, and air conditioners as well as gas liquefiers and low-temperature physics research tools.

1.2 Classification

Several completely different types of air engines were marketed in substantial numbers and have eventually become obsolete during the last 150 years. They all fulfilled a temporary need before other types of prime mover were developed. All these different models managed to produce at least a moderate amount of power, but failed to stand up to the competition of the modern internal combustion engine. Before starting a detailed description, an attempt is made here to classify* these various types. One purpose of such a classification is to define the term 'air engine', as used here, and to compare it with engines which are better known, such as, for example, gas turbines, whose working fluid consists almost entirely of atmospheric air and which may, therefore, also be considered to be a special kind of hot-air engine.

A logical division of air engines into three classes, in accordance with the fundamental principle used for the working cycle, was given by Slaby (1878). These classes were:

(1) Open cycle engines, where a fresh charge of working fluid is taken in for each cycle and where products of combustion mix with the working fluid.
(2) Open cycle engines, where a fresh charge of working fluid is taken in for each cycle, but with external heating.
(3) Closed cycle engines, where the working fluid is used over and over again.

In so far as gas turbines and two-stroke or four-stroke internal combustion engines can, strictly speaking, be regarded as varieties of hot-air engines, they belong to Slaby's first group. As a gross oversimplification, it may be said that the working cycle of hot-air engines of the first group usually resembles the 'constant-pressure cycle', those of the third group the 'constant-volume cycle', while those of the second group may be made to approximate either. In actual fact, these theoretical and, therefore, ideal thermodynamic cycles of reference bear only a very superficial resemblance to practical cases, but intelligently used they can serve a good purpose in allowing performance potential to be predicted and compared.

1.3 The regenerator

Although the three classes of air engines as defined above are entirely different from each other, throughout the period when the air engine enjoyed great popularity (i.e. during the latter part of the nineteenth century), models of each kind were in general use and engines operating according to each of the three

*A method of classification proposed by Rinia (1946) related only to one particular type of air engine, but was general enough to include such 'air engines' as conventional firearms, gas turbines, and rockets. A very ingenious, but rather artificial, classification of air engines was proposed by Grashof (1890) in accordance with the number of 'spaces' in which the thermodynamic cycle is performed.

systems continued to be built. Design was never concentrated on any one type to the exclusion of the others and, although a great variety of models were built, little notice was taken of successes or failures of earlier models. Indeed, many so-called 'improvements' were, in fact, just the reverse through lack of appreciation of the experience gained by earlier workers.

As will be seen later, Stirling's original invention of 1816 was in advance of practically every single design produced during the following 100 years. In order to appreciate this paradox, it is necessary to recapitulate the knowledge of thermodynamics in the early days of the hot-air engine, with special reference to the engine's most critical component, the regenerator.

When the first hot-air engines were built, the thermodynamic processes involved were a complete mystery. At that time heat was still considered to be that elusive 'caloric', a material and indestructible medium, which could be squeezed out of bodies like water from a sponge. While the first working air engine was built in 1807, Sadi Carnot's description (1824) of a thermodynamically reversible cycle did not appear until 17 years later. By the time the caloric theory was finally exploded when Joule established the mechanical equivalent of heat in 1849, engines of each of the three types had reached a stage of considerable technical development.

In order to understand fully the operational principles on which a hot-air engine depends, it is essential to grasp the functioning of the regenerator which was first invented by Robert Stirling in 1816 and applied by him to his hot-air engine (*Anon.* 1917a). Although a regenerator in itself is extremely simple, and a bundle of wires stuffed into the appropriate space is a fairly efficient arrangement, a thorough knowledge of thermodynamics is required in order to understand its functioning. To explain the action, it was described later as being similar to 'Jeffrey's Respirator' which was used at one time by consumptive persons to transfer the heat in the exhaled breath to the air inhaled, with allegedly beneficial effects to the patient.

The application of the regenerator to the air engine gave rise to an interesting controversy which could not be settled before the nature of heat was properly understood. It was thought by engineers of that period that in an air engine a certain amount of 'caloric' performed work by 'dilatation' (expansion) while being 'soaked up' in the regenerator. The common belief was that all of this 'caloric' could afterwards be recovered from the regenerator and go through the same process over and over again. The heat supply to air engines equipped with regenerators was thought to be necessary merely for making up losses. This is equivalent to the assertion that a hot-air engine could be made into a kind of *perpetuum mobile*, were it not for the fact that on repeated 'reconversion' of the caloric some of this precious substance is lost by 'degradation,' in other words, friction.

A great deal of ingenuity was expended in supporting such wrong theories. It may, in fact, be said that much effort in the development of hot-air engines was devoted to the pursuit of the most popular pastime of that period, the construction of a *perpetuum mobile*.

In the old literature inconsistent statements, based on a medieval conception of heat, can be found again and again. As a typical example, an English patent of Ericsson, in the name of Edward Dunn, stated as late as 26 December, 1850:

'. . . *the mode of applying the caloric being such that after having caused the expansion or dilatation which produces the motive power, the caloric is transferred to certain metallic substances, and again transferred from these substances to the acting medium at certain intervals* . . .'.

This evidently implies that according to the inventor's conception none of the heat had been converted into mechanical work, and all of it is available again and again to go through the same cycle.

After the mechanical theory of heat had become firmly established, such statements only discredited the principle of the hot-air engine in the eyes of scientists of that day. Cheverton (1852) gives an interesting account of this controversy, arguing as follows.

'. . . *caloric, in the mechanical view of the subject, is known simply as a force. Now* . . . *a force which, admitted to become for an instant extinct in one body, by transmission to another is in the next moment capable of becoming self-recruited* (is an) *assumption inconsistent with all natural phenomena. Yet the 'Caloric Engine' is chargeable with this absurdity so far as it is founded on the principle 'that the production of a mechanical force is unaccompanied by the loss of heat' and that 'caloric can operate over and over again'* (this amounts to) *affirming that power can be gratuitously exerted* . . . *in short that Newton's Third Law of Motion is untrue* . . .'

Even after the science of thermodynamics had reached a stage which resulted in the development of two-stroke and four-stroke engines, there were still some who decried the importance of regeneration for air engines, such as Zeuner (he recanted belatedly in the third (1907) edition of his famous textbook), and most hot-air engines in use worked without a regenerator.

It is recognized today that a regenerator is, in fact, the essential component of an air engine, and that the efficiency of a complete engine is largely determined by that of the regenerator. Stirling must have been aware of this fact and so was Ericsson, not by scientific reasoning, but probably by intuition. Unfortunately, Ericsson never succeeded in substantiating the statement made by him in 1855 that '. . . *we will show practically that bundles of wire are capable of exerting more force than shiploads of coal* . . .' It is rather sad to think that, in spite of this poetic truth, his only engine which achieved commercial success, and which was turned out in thousands, had no proper regenerator at all. Most of the later models of hot-air engines of all types were fitted with either no more than a vestigial trace of a regenerator or no provision for regeneration whatsoever.

The controversy about air engines and the way regeneration affects their output lasted a long time. When Faraday, who seems to have supported Ericsson in his claims, announced a lecture on air engines at the Royal Institution, many distinguished scientists came to hear him on this subject. Unfortunately Faraday was assailed by some doubt at the last minute and had to confess that he was unable to explain why Ericsson's engine managed to work at all (Church 1890).

Rankine (1854a) was the first to give a true explanation of the action of a regenerator in air engines. He had always been very interested in air engines and an excellent early theory is given in his famous treatise on the steam engine (Rankine 1908) It is nearly forgotten today that Rankine, together with Napier, designed an early air engine in which extended heat transfer surfaces were used for the first time [Patent No. 1416, dated 1853, by James Robert Napier and William John Macquorn Rankine (Bourne 1878, Knoke 1899)].

Thus, Stirling's invention of the regenerator was made nearly half a century before a proper understanding of its functioning became general. The first recognition of Stirling's genius in inventing the regenerator so many years in advance of any scientific knowledge of the subject came from Professor Fleeming Jenkin, who said (1883–84): '*. . . the regenerator is really one of the greatest triumphs of engineering invention.*'

After this discussion of regenerators, which has been necessary for putting the whole subject of air engines in its proper perspective, the design of individual engines that were in common use at one time can be explained. As a great number of entirely different engines appeared on, and disappeared from, the market during the nineteenth century, it will be better to treat separately the three systems outlined above in accordance with Slaby's classification, instead of describing the development of all types of air engines simultaneously in chronological order.

It will, therefore, be more convenient to describe first those hot-air engines that used fresh atmospheric air for each cycle and employed internal combustion, all of which are now obsolete, but excluding engines working on the four-stroke or two-stroke principle. After dealing with this group, which was doomed to failure from the beginning due to technical reasons, open-cycle machines with external combustion are dealt with. This system achieved a fair degree of popularity about 100 years ago, but its obsolescence for power production is also probably final, again on scientific grounds. The most promising scheme – namely, that of closed-cycle external combustion machines – is dealt with last.

Only engines which were once in common use, or which were forerunners of a type that later became popular, are described here. For further information the interested reader is referred to Knoke's excellent book (*op. cit.*) where further weird examples of hot-air engines of the late nineteenth century can be found, some of which had been constructed and were actually running, some of which had only been constructed, and some of which did not survive the drawing board stage.

1.4 Furnace gas engines

At the beginning of the nineteenth century the use of steam engines became more general. They were the only source of mechanical power that was available, other than water-wheels or windmills. Watt's introduction of condensers for steam engines had already considerably improved their power and economy, and the next step forward was the use of higher steam pressure. Although Watt was very much against this and preferred to use engines with steam pressures no higher than about 5 lb/in^2 gauge, this innovation was duly made by Trevithick with a consequent further increase in specific power. This led to a very high accident rate, because boilers constructed from the materials available at that time were liable to burst at the least provocation with attendant loss of life or severe injury.

It was inevitable that, on the grounds of safety alone, many engineers were tempted to substitute air for steam as the working fluid for existing engines. In principle, such a modification merely means that a compression cylinder must be used instead of a feed pump, and that the products of combustion from the fire must be admitted direct to the cylinder without the intervention of a boiler. The reciprocating steam engine, including valve gear, could then be used practically unchanged, but the effect of hot gases was extremely harmful to the engine and indeed would have wrecked it fairly soon.

The very first engine working on this principle, of which records exist, was probably that designed by Sir George Cayley, Sixth Baronet (Cayley 1807). He was an inventive genius who carried out a large number of scientific experiments at Brampton Hall, near Scarborough, and whose investigations ranged over a wide field. Among his notable innovations were caterpillar tractors, pneumatic railway buffers, automatic brakes, streamlining, safety belts, and wheel spokes under tension. In a recent talk on the BBC he was even credited with the invention in 1799 of the fixed-wing aeroplane. He was the first to construct a rigid aeroplane which actually rose off the ground carrying a passenger. Some flight tests were made with a boy on board, others with one of his servants. On one occasion, in 1852, after a successful flight with his reluctant coachman as test pilot, that brave pioneer is reported to have given his notice, shouting *'I was hired to drive, not fly!'* (Gibbs-Smith 1957).

Unfortunately, detailed descriptions and illustrations of many of Cayley's inventions failed to survive, so the engine of Fig. 1.1 is important as a record of what was probably the first practical hot-air engine ever built. Cayley developed air-operated prime movers over many years and in 1837 took out a patent for an improved engine, which was manufactured by the Caloric Engine Company (Poingdestre 1849). Air was compressed in a large cylinder and passed under pressure through a furnace as the combustion air. The products of combustion were then admitted to an expansion cylinder which was practically identical to that of a steam engine of that period, from which the work output was derived. Though this system was very crude, the difference between the work performed in the expansion cylinder and that consumed in the compression cylinder was just sufficient to produce a small amount of power. For his latest engines, 5 b.h.p. with a consumption of 20–22 lb of coke per hour were claimed.

After Cayley, a number of different designs for such engines were brought out, which varied greatly in their degree of complexity. Constructional features are adequately illustrated by showing two extreme cases, a 'Gradual Combustion Engine', in Fig. 1.2a, and a 'Furnace Gas Engine', in Fig. 1.2b. The former shows the most simple and compact arrangement that was achieved with this principle, while the latter is an intricate and complicated piece of machinery. Although both types appear to have functioned fairly well, they were not based on sound engineering principles and were therefore doomed to failure as a practical proposition.

Figure 1.2a represents a fairly advanced form of this type of engine and is reproduced from a book entitled *Modern Engines and Power Generators* (Kennedy 1904–5). A double-acting piston was used with different effective areas on the two sides due to the thickness of the piston rod. Valve *A* is an automatically operated inlet valve, through which all of the air supply for the fire is sucked in. The grate *F* formed an integral part of the cylinder casting. Valves *D* and *E* were operated by linkwork which is not shown. Valve *D* served to admit the hot combustion gases to the cylinder, while *E* was the exhaust valve.

When the piston moves from right to left, *D* closes and *E* opens, so that the gases present to the left of the piston are expelled. The pressure to the right of the piston and in the grate drops below atmospheric pressure due to the increase in volume, so that *A* opens and the space to the right of the piston is filled with fresh air from the atmosphere. As the pressure on the left of the piston is slightly above, and that on the right slightly below, atmospheric pressure, the energy required for this stroke is derived from the flywheel. At the end of this

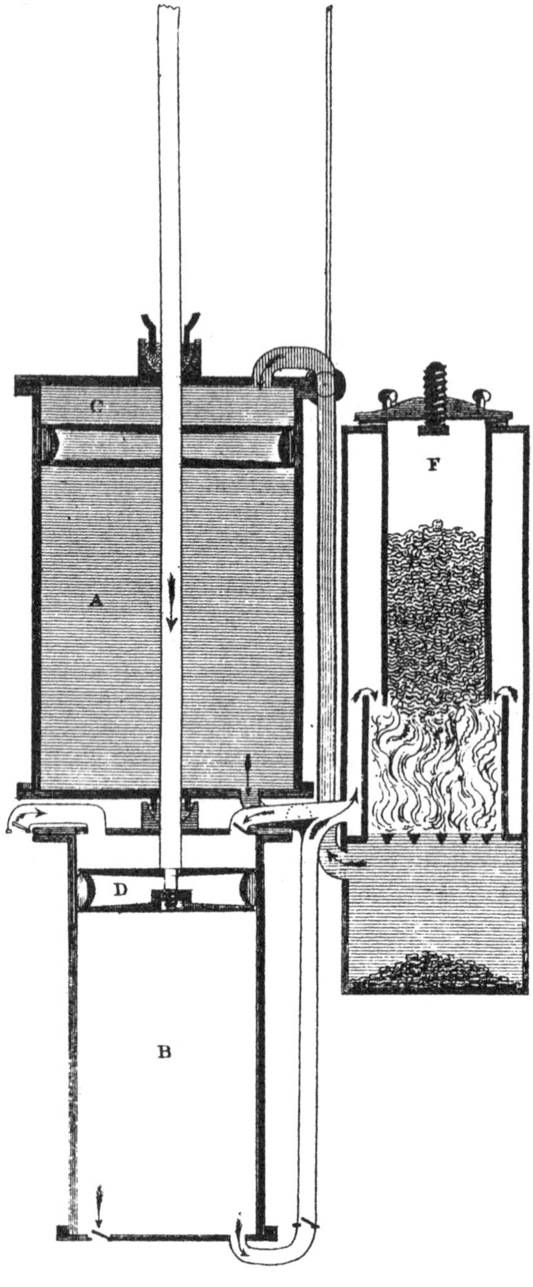

Fig. 1.1 Early continuous combustion air engine. After Cayley (1807)

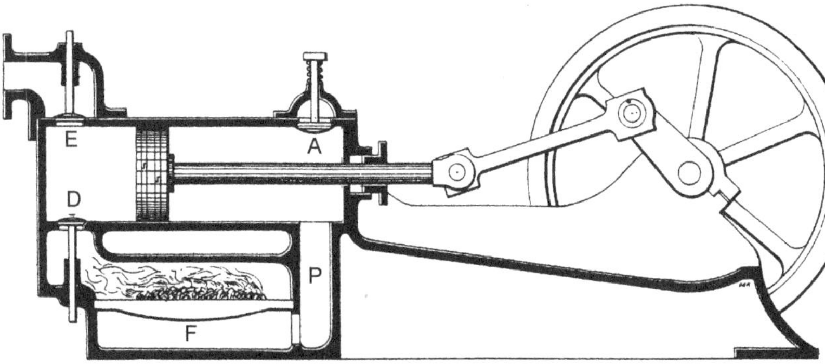

Fig. 1.2a A 'gradual combustion engine' with double-acting piston and integral furnace and cylinder assembly. Built in 1880 and used for water pumping. After Kennedy (1904–5)

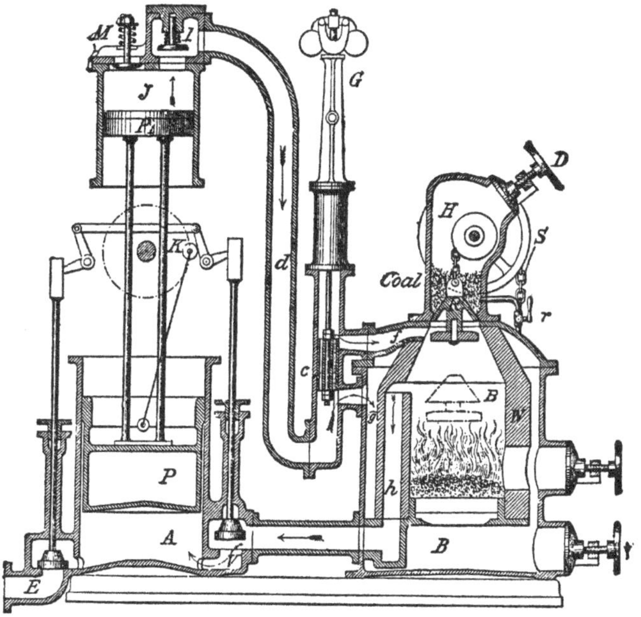

Fig. 1.2b A Buckett 'furnace gas engine' with separate cylinder assembly and furnace. This engine achieved considerable popularity between 1875 and 1885. After Donkin (1911)

stroke, *E* closes and *D* is opened, and as the piston starts on its return stroke some of the fresh air comes into contact with the coal in the grate *F*, so that heat is imparted to it, which raises the pressure. This pressure, of course, acts on both sides of the piston. However, there is a net force equal to the area of the thick piston rod multiplied by the pressure, and expansion is accommodated by driving the rod out of the cylinder.

As the furnace of firebox *F* is placed inside the main cylinder assembly, while at least some heat is transferred by conduction when valve *D* closes, one might say that this engine combines features both of external and internal combustion engines and is, therefore, a hybrid between these two types. It is interesting as an engineering curiosity, but as a power producer it probably manages to combine in one model most of the disadvantages that any hot-air engine could and did have. Kennedy (*op. cit.*) stated that: '. . . *as a hot-air engine with reciprocating machinery, it is of little present value, except for pumping water in country districts, where its extreme simplicity is a great consideration* . . .'.

The second engine described here, and shown in Fig. 1.2b, is the Buckett engine, an improved form of Cayley's design, which was built by the Caloric Engine Company round about 1880 (*Anon.* 1883). The two main components of this plant are the furnace, shown on the right, and the cylinders, on the left. The power piston *P* working in the expansion cylinder *A*, and the compression piston P_1 working in the compression cylinder *J* are joined by two rods. Valves *M* and *I* in cylinder *J* are automatically operated as in some air compressors, while valves *V* and *E* in cylinder *A* are operated by linkwork connected to the crankshaft, as shown diagrammatically in the illustration. Valves *E* and *V* are of a special shape, in order to withstand the effects of hot gases, which is intermediate between that used for early steam engines and for poppet valves in modern internal combustion engines. The compressed air leaves *J* on the upstroke and is divided by a slide valve *c* into two streams, one providing the primary air for combustion at *g* and the other the secondary air at *f*. This slide valve *c* is operated automatically by the governor *G*. The primary air stream *g* is passed round the hot brickwork for preheating and is then admitted to the furnace from underneath the grate *B*. The secondary air is necessary in order to limit the temperature of the air reaching the working cylinder, a practice perpetuated in today's gas turbine. The products of combustion enter the cylinder *A* through the inlet valve *V* and, after having delivered work during the upstroke of the piston *P*, leave the cylinder *A* through the exhaust valve *E*. There are several refinements, such as the hopper *H* which allows stoking without stopping the engine, and a special arrangement for cooling the inlet valve *V* by means of cold air.

Quite a number of such engines were in use at one time and a few test data were published. For example, one engine of similar construction was designed by Mazeline to run a paper mill. This had a cylinder diameter of 1 m and a stroke of 1.5 m, but according to Tresca (1867), its performance was not very good. Fleeming Jenkin (*op. cit.*) quotes a fuel consumption of 1.8 lb of gas coke per indicated horsepower-hour and 2.54 lb per brake horsepower-hour, which corresponds to a thermal efficiency of 7.9 per cent for this type of engine. Some test results are also given in *Engineering* (1875) and by Donkin (1911). It was stated that the indicator diagrams obtained from this engine had a similar shape to those from steam engines.

Benier, whose name is well known from the development of the gas engine, constructed a similar engine which was approved by the Trinity House authorities in 1886 for the warning systems on headlands. Other earlier engines, such as Hook's engine and A. and F. Brown's of New York, also differed in minor detail. For example, in Brown's engine, the cylinders were separate instead of coaxial (Knoke *op. cit.*).

Several other engines using a similar working principle to the two types described here were proposed. For example, an engine (Patent No. 3807, Söderström 1868) used a single cylinder in direct communication with the furnace and a piston working as compressor on the downstroke, and expander on the upstroke. Bourne (*op. cit.*) gives an excellent illustration and description of this ingenious engine. Other engines with internal combustion are described in *Engineering* (1875), by Donkin (*op. cit.*), and by Kooke (1873).

Although a regenerator could have easily been applied to all engines of this type, this was never done, probably because it was difficult enough to keep the temperature of the working cylinder down to a safe value without regeneration. The dust and ashes carried over into the cylinder from the fire, combined with the difficulty of maintaining at least some lubrication in the power cylinder, destroyed the engine fairly rapidly in nearly all designs.

It is quite apparent now that this system is entirely obsolete and its only interest is in being the forerunner of the later gas engine where combustion takes place inside the cylinder rather than in a separate furnace, but where the high temperatures occur only for very brief instants.

1.5 Ericsson engines

Various engines belonging to the first and third groups were developed by a great number of different inventors. On the other hand, one man alone, John Ericsson, believed in the system of open cycles with external heating, nearly all known engine types of this second group being originally devised and perfected by him. He built a large number of experimental engines and, in due course, evolved two basic designs.

John Ericsson was born in Vermeland, Sweden, but spent most of his professional life in England and America. He became famous for his many inventions and innovations in connection with locomotives, steam fire engines, screw propellers, ships, torpedoes, etc. His first vessel of the *Monitor* series was the forerunner of all later iron-clad turreted warships with machinery below the waterline. He designed and built it in about 100 days, and its appearance at a critical moment in the American Civil War reversed the fortunes of the Northern States' Navy.

Ericsson had conceived the idea of an air engine in his early youth and devoted many years, both in England and in America, to its development. Shortly after his arrival in England in 1826, he commenced experiments and even unsuccessfully opposed Stirling's patent for a regenerator (Church 1890). His first patent, No. 5398 of 1826, referred to an engine which had a compressor cylinder delivering pressurized air to a hermetically sealed furnace. The products of combustion were passed through a vessel fitted with shelves over which water was made to trickle. The mixture of steam and hot gases was then used to drive his engine (Bourne *op. cit.*).

Ericsson also built an air engine in 1833 which, as far as can be ascertained today, used a closed cycle with constant-pressure heating and cooling. No detailed records of this construction have survived, but it seems (*Anon.* 1834*a* and *b*) that the working principle was similar to that of the constant-pressure cycle, as applied by Joule in a reversed form of refrigeration.

Though Ericsson himself was not very satisfied with the results, it caused a great sensation when it was exhibited and even evoked the following lyrical remarks from Sir Richard Phillips (1833):

'. . *with a handful of fuel applied to the very sensitive medium of atmospheric air and a most ingenious disposition of its differential powers, he beheld a resulting action in narrow compass, capable of extension to as great forces as ever can be wielded or used by men . . .*'.

During Ericsson's stay in England he experimented not only with this system but also with several other designs which used closed cycles. For some unknown reason Ericsson soon abandoned all these more promising closed engine cycles and began to concentrate his efforts on a system that had very little promise of success for practical reasons, namely, open cycles with external heating. In spite of many setbacks he managed to bring this system to a high degree of perfection. Ericsson moved to America in 1839 where he continued to work on the development of air engines, as well as on his many other innovations. Between 1840 and 1850 he built eight experimental engines, with some improvement over the preceding design added in each model. The last one seems to have been quite a success. It developed 5 h.p.

and was installed in the Delameter iron foundry in New York. A year later a 60 h.p. engine was built which was also shown at the great London Exhibition of 1851 [Poppe (1853)].

The working principle of Ericsson's engines is rather difficult to understand and will, therefore, be explained in detail by reference to Fig. 1.3 which carries Ericsson's own annotation. In the general layout there is a lot of similarity between this engine and the Buckett engine shown in the preceding figure. A compressor and expansion cylinder with coupled pistons *c* and *c*', and similar valve gear are used in this engine as well, but there the resemblance ends. As it is an external rather than an internal combustion engine, Ericsson's construction provides for complete separation between working fluid and products of combustion, relocates the valves into the working fluid circuit, and employs a regenerator in an ingenious manner.

Referring to Fig. 1.3, cylinder *d* is built into a furnace with its bottom in contact with the fire. The compression piston *c* is coupled mechanically with the power piston *c*', but is working in a separate cylinder *b*. These pistons are connected to a crankshaft with a flywheel in a conventional manner. A regenerator *f* and a large air receiver *a* can be seen in the diagram. Valves at the top of cylinder *b* again correspond to the automatic valves in an air compressor, while valves *g* and *b* are constructed similarly to those employed in steam engines and are actuated by an eccentric on the crankshaft which is not shown. The operation of this engine is described below.

The receiver *a* is charged up to a constant elevated pressure by means of piston *c*. When both pistons are at bottom dead-centre position, valve *g* is open and valve *b* closed. Air from receiver *a* now starts to pass through the regenerator, where heat is imparted to it. It is further heated below piston *c*' by contact with the hot cylinder bottom, and with the lower part of the piston (which was usually filled with brick dust). This helps to drive the pistons up, but when half the stroke is completed, valve *g* cuts off the supply, and further expansion is intended to be isothermal. Valve *b* opens on the downstroke, while valve *g* remains closed, and the working air is expelled through the regenerator, where much of its heat is absorbed. It will be readily conceded that the design of this engine is ingenious, especially with regard to the application of the regenerator (*Anon.* 1860e, Grashof 1860, Röntgen 1888).

Ericsson must have been too confident about the success of such an engine, because his next enterprise in the course of the development of hot-air engines amounted to one of the most spectacular failures in the history of engineering. This was mainly due to the fact that Ericsson failed to appreciate that a good working scheme in one size is not necessarily as good a proposition when the machinery is greatly enlarged.

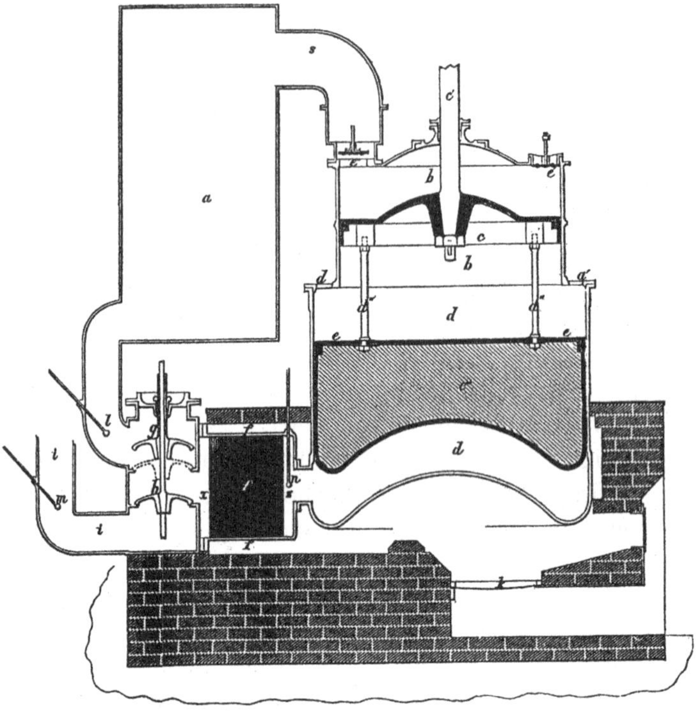

Fig. 1.3 Experimental 'caloric' engine designed by Ericsson and built in New York in 1851. Note the separate compressor–displacer cylinder and externally mounted pressure vessel and regenerator. After Ericsson (1876)

A company for the exploitation of this invention for the propulsion of naval vessels was formed in America with John B. Kitching at the head, with the support of the Secretary of the Navy, Kennedy. This company immediately started to build the 2200-ton vessel 'Ericsson', which was propelled by two paddle wheels and which was to be powered by four caloric engines of Ericsson's design. In due course, this vessel made its maiden voyage on 15 February 1853. The engines fitted were all similar to that shown in the diagram (Fig. 1.3), except that two regenerators, placed side by side and working alternately, were provided for each cylinder.

An idea may be gained of the unwieldiness of the power plant when it is considered that the cylinders were no less than 14 ft in diameter with a stroke of 6 ft. With the manufacturing methods then known, any attempts to make the pistons close fitting must have resulted in enormous friction losses and it must have been a veritable nightmare to maintain these vast cylinders reasonably leak-proof. The large size must also have had a bad effect upon the heat transfer due to the very unfavourable surface–volume ratio; otherwise, the engines were of advanced design. The four regenerator assemblies that were provided were made of a network of wires, whose combined length was nearly 50 miles, and a weight of 33 000 lb, which helped to heat the air in this engine to a maximum temperature of only 195 °C.

Ericsson's design calculations indicated a power output of 600 h.p. and a fuel consumption of 8 tons of coal per 24 h, or 1.11 lb per horsepower-hour. Unfortunately, these figures were later proved to have been wildly optimistic. It was stated that on the trial trip only 300 h.p. was attained with a fuel consumption of 2.2 lb of anthracite per horsepower-hour. It should not be thought, however, that these results were too disappointing, because marine engines of that day still required 3.11 lb of coal per horsepower-hour, so that in actual fact these figures showed a definite saving in fuel. The consumption quoted per indicated horsepower-hour was 1.87 lb, but this is evidently wrong because it would correspond to a mechanical efficiency of 85 per cent, a figure surprisingly high for machinery in those days. It is obvious today that the main drawback of these engines was their enormous weight and bulk. The maximum speed of the engine was only a ponderous 9 r/min, while the two 32 ft diameter paddle wheels propelled the vessel at about 6.5 knots (Ferguson 1961).

At first Ericsson thought that performance could be improved by modifications to the machinery. He therefore completely redesigned the engine, substituting two cylinders of smaller diameter and longer stroke for the four originals. This new plant proved even less efficient than the original, and early in 1854 was unceremoniously removed from the ship and replaced by a set of steam engines. Even this did not improve the vessel's bad luck, as it went to the sea's bottom in 1855 (*Anon.* 1855, Ericsson 1855).

It may be thought that after this fiasco Ericsson might have given up the idea of a 'caloric engine', but he still continued his experiments. After building about 15 further test machines, he perfected and put into production an engine which worked according to the same cycle, but where the mechanical details were entirely different. The whole process was now taking place in one single cylinder. This engine had the distinction of being the first 'mass-produced' model of any hot-air engine. It proved so successful that by 1860 about 3000 had been sold and were in regular use in many countries, including the United States, England, Germany, France, and Sweden.

The appearance of this engine is shown in Fig. 1.4, which is a photograph of an engine preserved in the Science Museum, London. Ericsson's own drawing is reproduced in Fig. 1.5. The main cylinder was closed at one end by the power piston *A*. Inside, a feed – or transfer – piston *F* was actuated by a rod passing through the power piston. The other end of the cylinder was closed by a firebox

Fig. 1.4 Industrial Ericsson engine in its most advanced form, built in 1869. Reproduced by courtesy of The Science Museum, London

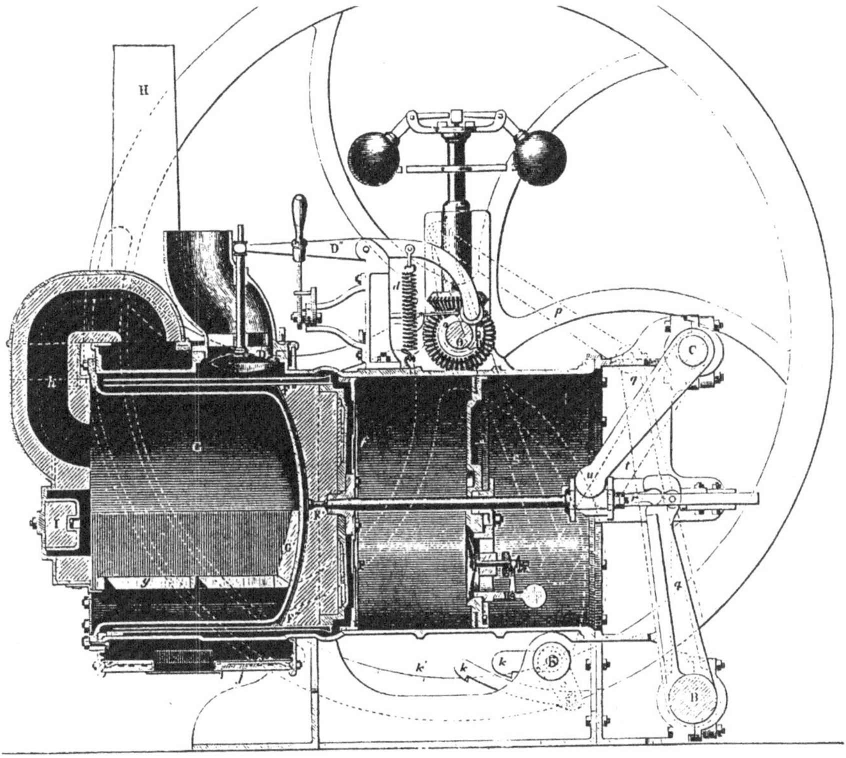

Fig. 1.5 Cross-section through an Ericsson engine similar to that of the previous illustration. Several thousands of this engine were built in America and Europe between 1855 and 1860. After Ericsson (1876)

G complete with grate *g* and ash pit, in which fuel was burnt, and which projected into the cylinder for some distance. The left end of the cylinder was therefore heated from inside, as well as from the outside, by passing the products of combustion from the fire into an annular heating jacket, which surrounded the cylinder, before exhausting through the smoke stack *H*. In order to increase the transfer of heat from the fire to the working fluid, two cylindrical metal shields were fitted. One of these surrounded the firebox and was fixed to the cylinder, and the other shield was attached to the feed piston and reciprocated in the annular space between the firebox and the other shield. The working fluid passed through sets of air-operated valves in the two pistons on their outstrokes and was rejected to atmosphere after having gone through valve *D* which was actuated by linkwork. To prevent direct conduction of heat between the hot space and the cold space, part of the transfer piston was made hollow, and the space filled with pulverized charcoal. The action of the pistons was such that they moved out of phase and with different amplitudes, driven by the system of cranks and levers visible in the photograph. One half of the flywheel was made heavier than the other by casting cavities around the rim, half of which were filled by casting lead into them in order to assist with dynamic balancing and facilitate starting. A centrifugal governor opened a small release cock to atmosphere whenever the speed rose to an excessive value.

The sequence of events in an operational cycle of this engine is as follows: starting with both pistons near their outer dead-centres, valve *D* is open so that the hot gas trapped in the cylinder to the left of the feed piston can escape to atmosphere. The feed piston *A* is travelling to the left, which assists in scavenging the cylinder, whose contents at this stage are slightly above atmospheric pressure. The valves in the feed piston *F* are, or course, closed, and the increase in volume between the two pistons opens the valves in the outer piston, so that cold air from outside begins to enter the space between the two pistons. After the feed piston has travelled a short distance, the power piston also commences its stroke to the left, but as the amplitude of the latter is smaller, and as there is an appreciable phase difference between these movements, the volume between the pistons increases continually until the feed piston has reached its inner dead-centre. The feed piston then returns, which closes the valves in the power piston, while the valves in the feed piston open. The exhaust valve *D* closes when piston *A* is at inner dead-centre. The working piston continues to travel to the left for a short distance, which compresses the enclosed air slightly. A further pressure rise occurs when the cold air passes through the valves in the feed piston, where the temperature is raised through contact with the hot metal plates. Mechanical work is then delivered to the crankshaft on the outstroke of the power piston by virtue of the expansion of the hot air. During this outstroke of the power piston the feed piston is travelling to the right as well, approaching the power piston continually, so that further quantities of cold air are made available for heating.

It should be noted that though this engine is not provided with a proper regenerator, some of its components function in a way resembling the operation of a regenerator by abstracting heat from the hot exhaust gases and storing it until this heat is transferred to a new charge taken in. The valves in the transfer piston *F* are positioned in such a manner that air on entering the hot space to the left of *F* is made to flow between the outside of the inner shield and the inside of the outer shield, and then between the inside of the inner shield and the walls of the firebox. By these means some heat can be abstracted from the two shields by incoming air. When the cycle is completed, the air is exhausted through these same passages in the reverse direction so that some of the heat is absorbed by the two shields and stored in readiness for the next incoming charge. In this respect this engine is similar to the Lehmann engine in the third group, which will be described later, where also no regenerator was provided but where these functions were partly performed by different components which, in this case, were the walls of the transfer piston in conjunction with the cylinder walls (Bourne *op. cit.*).

It will be appreciated that these engines were very solidly built and were correspondingly heavy. Warming up was rather slow and it took about 2 h to get one started. A later improvement by Brown of New York reduced this to about 20 min. The long and heavy exposed levers and the crude method for actuating valves by means of overhung weights made these engines exceptionally noisy and they worked with a deafening clatter not usually associated with air engines. Their rating was approximately 1/2 h.p., while their mechanical efficiency was commonly stated to be 40 per cent. These engines were somewhat extravagant in fuel consumption and E. H. Schmidt (1861), who performed some tests, quotes 15 lb of coal per brake horsepower-hour.

One very successful application of Ericsson's engine was for a coast fog signal. Dayboll, the inventor of the first practical fog trumpet, installed such a warning signal on Beaver Trail Island, Providence, which was worked by a donkey. As the fog sometimes lasted longer than the donkey could cope with, Ericsson suggested to him that this engine be used. That particular engine was later tested by Faraday in 1862. Its power output was about 3/4 h.p. and it was still doing duty on a lightship after 30 years' service – an early demonstration of the long life and reliability of air engines (Shaw 1880, Matschoss 1925).

2

The Stirling engine

2.1 The invention

In open-cycle engines mechanical work is performed by a quantity of working fluid which is made to undergo successively certain operations, such as induction, compression, heating, expansion, and exhaust. On the other hand, in closed-cycle engines these phases merge into each other, and while part of the working fluid may be heated in one part of the engine, in another part it may simultaneously be cooled. Thus the process is rather more difficult to understand. The invention of the closed-cycle external combustion engine by Stirling in 1816 is therefore probably one of the most amazing innovations that has ever been made. It was so much in advance of scientific knowledge at the time that at least 30 years passed before anyone was in a position to understand what made the engine work at all. Nearly 100 years later engines were still 'invented', which, due to lack of understanding of the fundamental principles, missed such essential features as the provision of a regenerator, or compact cylinder arrangement, whose importance Stirling must have realized intuitively.

This invention was made by the Reverend Robert Stirling, D.D., of Cloag, Methvin, Perthshire, when he was 26 years old and had just been ordained to his first parish. Robert Stirling's family produced a number of prominent engineers over several generations, from his grandfather, Michael Stirling, inventor of the first rotary threshing machine in 1756, to his four sons, all of whom became well known, mainly as railway builders, in places as far apart as Honolulu, Chile, and Scotland.

The original patent, No. 4081 of 1816, had the obscure title *'Improvements for Diminishing the Consumption of Fuel, and in particular an Engine capable of being Applied to the Moving (of) Machinery on a Principle Entirely New'*. In it Stirling not only described the construction and use of a regenerator for the first time in history, but also foresaw its principal applications, such as for glass furnaces or iron smelting. It also included a description of the first closed-cycle hot-air engine, as shown in Fig. 2.1. This is reproduced, with minor corrections, from the original patent specification (London version). It should be noted that this engine was considered by Stirling to be a logical development based on the regenerator, a fact sadly neglected by later innovators.

Even today, the working principle of this original Stirling engine – as that of many later engines based on this prototype – is obscure. Diagrams explaining the working principle are therefore shown here in Fig. 2.2. On the right, an

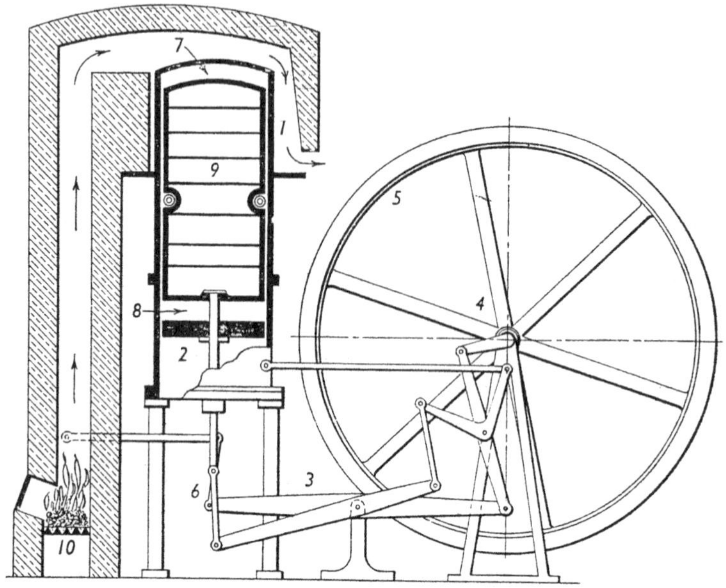

Fig. 2.1 Reproduction of drawing showing the first Stirling engine from the patent specification of 1816 (London version). The engine used in 1818 for pumping water from a quarry (*Anon.* 1917a) is widely supposed to have looked like this

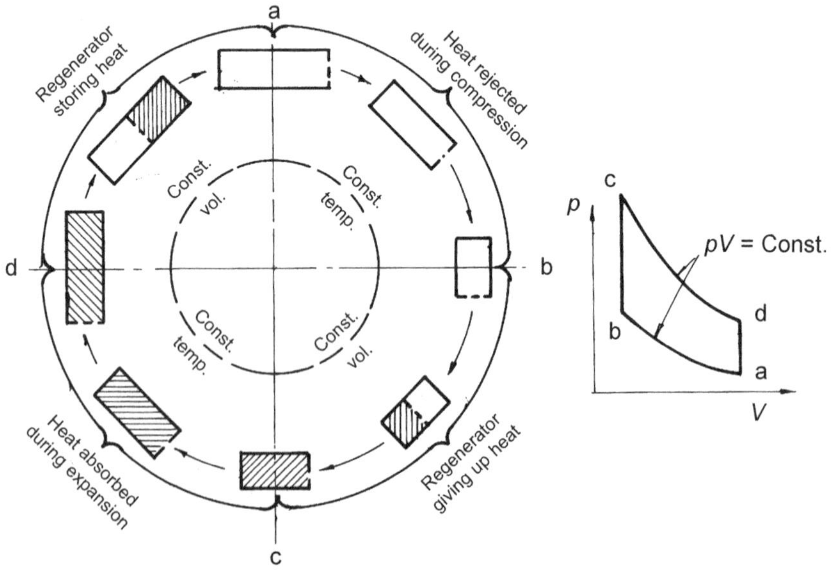

Fig. 2.2 The principle of operation for the Stirling cycle is shown here with reference to an indicator (*p*–*V*) diagram

idealized 'Stirling cycle' is drawn as a pressure–volume (indicator) diagram. (It is not possible to construct a unique temperature–entropy (*T–s*) diagram for any conceivable practical embodiment.) Terminal phase points are marked *a*, *b*, *c*, and *d* respectively. On the circular phase diagram, each of the four quadrants corresponds to one of the main phases. The enclosed working fluid is represented as a rectangle whose area corresponds to the variable volume, with the position of the regenerator indicated by a dotted line. The temperature of the working fluid is indicated by cross-hatching that portion which is at the higher value. The sequence of events is clockwise, as indicated by arrows. At phase point *a* the volume of the *low*-temperature fluid is a maximum. Between *a* and *b* this temperature remains constant, while the volume decreases up to point *b*. Between *b* and *c* the volume is kept constant and the regenerator passed through the working fluid, or conversely, the working fluid through the regenerator. In either case, the pressure of the working fluid rises uniformly throughout. After point *c* is reached, volume increases. The expansion takes place at a generally elevated temperature, giving a surplus of work over the preceding (low-temperature) compression. The regenerator now returns to its starting point (*d* to *a*) with overall volume constant, so that no work is involved. Thus a Stirling engine is a system whereby the total volume of a gas is periodically altered by means of one or more pistons, while successive transfers of the working fluid between two spaces at different temperatures is via a regenerator.

2.2 Working principle

The operation of the Stirling engine shown in Fig. 2.1 can now be explained with reference to this cycle diagram. The working fluid was confined in a vertical cylinder 1 about 10 ft high. Volume was varied in accordance with the diagrams of Fig. 2.2 by means of a piston 2, about 2 ft in diameter. This piston was driven by a mechanism usual for steam engines of that time, that is, by means of a rocking beam 3 worked by the crankshaft 4, with the large flywheel 5, about 10 ft in diameter, mounted on it. As accurate sliding surfaces for a crosshead could not be produced with the materials and workmanship available in 1816, without *'untowardly frictions'* in the words of the period, the piston is driven via a link motion 6, corresponding to Watt's straight-line mechanism. The only novel feature of this mechanism is that it was turned upside-down in order to keep the operating temperature of the piston at a low value. The working fluid enclosed in the cylinder 1 was divided by the displacer 9 into two parts – the hot space and the cold space, marked 7 and 8 respectively. The upper space 7 is kept hot by products of combustion from a fire in grate 10. The gases are led up a duct and surround the upper part of the cylinder before being passed up the smoke stack. While this jacketing by hot flue gases kept the upper end of the cylinder hot, the lower end was kept cool.

An arrangement for water cooling the lower space 8 is mentioned in the specifications, but is not shown in the illustration. This is because an engine of this type probably works better without this water cooling, a rather paradoxical fact that can be attributed to excessive heat conduction along the cylinder and which has been confirmed by the writer in experiments with a similar engine. The displacer 9 had the regenerator mounted in a narrow annular space on its cylindrical surface and was reciprocated through the mass of the working fluid in accordance with the principle shown in Fig. 2.2. The link motion actuating it consisted of a central push rod passing through piston 2, and an auxiliary beam worked out of phase to the main beam by a cranked lever mounted on the connecting rod. There was probably a slider on the parallel motion, though the illustration is not clear on this point. The displacer was hollow, made of thin riveted iron sheets, and had a diameter rather smaller than the cylinder. It was fitted with small wheels to keep it central, which left an annular space about 0.5 in thick for the regenerator. This regenerator was made of thin wire, which apparently was wound on the displacer body in a spiral fashion with successive layers criss-crossing at right angles until the annular space was nearly filled. The engine was also fitted with a governor which opened a cock near the lower part of the cold space to atmosphere whenever the engine was racing on light load.

Although figures for power output of engines dating back to the early nineteenth century are hardly reliable, by some accounts the engine shown in Fig. 2.1 produced about 2 h.p. An early model was installed in 1818 at an Ayrshire quarry for pumping water and worked for about 2 years until the cylinder cover became overheated. It is quite likely that this may have been the first practical application of all air engines (Manby 1853).

2.3 The patent

A romantic story was connected with the London patent specification (*Anon.* 1917*a*, *b*, *c*, Prosses 1917) when the specification was reprinted in honour of the centenary of the invention. Apparently it was customary in 1816 not only to 'seal' an invention but also to 'enrol' it within a stipulated period, usually about 6 months. It was necessary to take the specification to one of three enrolment offices in London, and an extra payment of £5 had to be made. This fee was additional to many other payments that were needed in those days for the issue of a patent, and for this the specification was copied on to parchment and 'enrolled' by stitching this new section on to the end of the preceding patent to make one long continuous roll. These ancient rolls are still stored at the Record Office, but it would appear that Stirling's specification is not included. Five pounds must have been a lot of money in those days, and it would not be surprising if at the time Stirling was not able to afford such a large sum in addition to fees for an agent in London or his own travelling expenses, and therefore omitted 'enrolment'.

Thus it would appear that Stirling's original English Letters Patent had become null and void, and although the title was entered in the ancient 'Docquet Book of the Great Seal', in which patents granted up to 1852 were registered, the subject matter and the drawings intended for London remained lost for the next 100 years. In a very thorough survey in 1884, made by one of the foremost authorities on heat engines in a series of lectures on caloric engines (Jenkin 1885), no mention of the original specification is made. Though the author knew that Stirling had invented the regenerator and certain types of air engines, he did not know about either of Stirling's specifications (Edinburgh or London) of 1816 or of any engine constructed in accordance with the layout shown in Fig. 2.1.

The way in which the missing documents eventually turned up is still shrouded in mystery. Apparently some papers were cleared up in an unspecified private house, and, lo and behold, Stirling's original specification was discovered and duly returned to a member of his family. Perhaps a researcher with an historical rather than a technical interest in this matter will one day shed further light on the circumstances of the mysterious disappearance and eventual recovery.

It seems that several people later tried to contest Stirling's patent for the regenerator. Among these was Ericsson, who, as mentioned before, unsuccessfully claimed priority for a regenerator (Church 1890). During the better-known Neilson hot-blast trial concerning iron smelting, copies of the Scottish patent were produced which had been duly registered independently from the English counterpart (*Anon.* 1917a, c).

Further development of the original model shown in Fig. 2.1 was carried out by Robert Stirling in collaboration with his brother, James, who was an engineer by profession. One early improvement which James Stirling introduced was the use of an elevated pressure level by means of a force pump. Another innovation of rather doubtful value was the construction of a 'twin engine', which had a double-acting power piston and separate displacer cylinders which, due to the loss of Stirling's London patent, was later erroneously thought to be the forerunner of all closed-cycle air engines.

One of the very rare illustrations of such an engine that has survived is reproduced in Fig. 2.3 from a woodcut (Galloway 1830). The huge flywheel (16 ft diameter), the ornamented beam pillar, and what corresponds to an eccentric strap and rod, look similar to parts of old steam engines. It is quite likely that, in fact, many parts of the transmission mechanism had been adapted from a contemporary beam engine. Experts in engineering history will note the usual combination of a Watts straight-line motion for driving the link corresponding to the condenser pump rod with a pantograph mechanism connecting with the main piston rod, an arrangement that had become conventional for most beam engines since its introduction by Watt many years earlier.

In this engine (Fig. 2.3) two large displacer cylinders are provided, as shown in the left foreground. Each of these must have been heated by a fire underneath, while a displacer–regenerator reciprocated inside. These displacers were provided with four guide rods and were driven through an auxiliary rocking beam, so that they moved 180° out of phase with each other. From each of these large cylinders, short connecting pipes, rather picturesquely called 'nosles' in contemporary descriptions, led to the main double-acting power piston. Thus this engine worked on two cycles, carried out respectively in each of the displacer cylinders in conjuction with half the power cylinder, either above or below the double-acting piston. A fourth, smaller cylinder is fitted between the power cylinder and the main beam pillar, in a position usually occupied by the condenser pump in steam engines. This is an air pump which was used to compress the air from atmosphere into the two working spaces.

In the old engraving, reproduced here without substantial changes, several obvious mistakes had been made by the artist. The displacer linkage driving the small rocking beam could obviously not function as shown, but would simply lock. The suspension of the air pump rod is also incorrectly shown – in practice there would have been a joint halfway up the vertical link. A diagrammatic distorted section through this and similar later engines built by Stirling is, therefore, shown in Fig. 2.4, in which an attempt was made to reconstruct the linkage that was probably used in these engines, and in which the layout of the cylinders and the working spaces is clearly shown.

Between 1824 and 1840 several improvements and modifications were made by the Stirling brothers to their earlier designs, and further patents were registered in their joint names in 1827 and 1840 (Nos 5456 and 8652). Figures 2.5 and 2.6 have been reproduced from these later specifications. In Figure 2.5 a section through the furnace, displacer cylinder assembly, and power piston in an earlier improvement is shown. The regenerator, which in this design was made of cylindrical metal sheets, is here transferred into an annular space surrounding the displacer, so that instead of being carried through the air by the displacer, the air is pushed through this stationary regenerator. The displacer was made of a hollow metal casting which was filled with brick dust. For some reason this was widely copied later, and some magical properties seem to have been ascribed to brick dust. Though many later hot-air engines dispensed altogether with a regenerator, they still used brick dust for filling their displacers.

In experiments with a machine built in 1824, Stirling noticed that although

Fig. 2.3 Old engraving showing a beam engine working as a two-cycle Stirling engine, dating back to about 1827. After Galloway (1830)

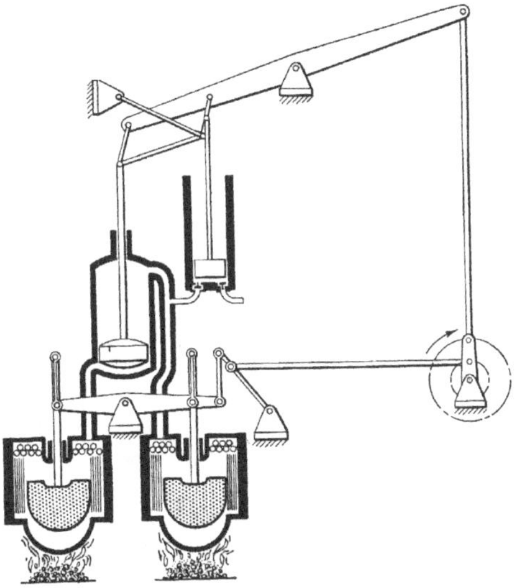

Fig. 2.4 Diagrammatic cross-section through an early two-cycle Stirling engine showing the arrangement of the driving mechanism that was probably used (not to scale)

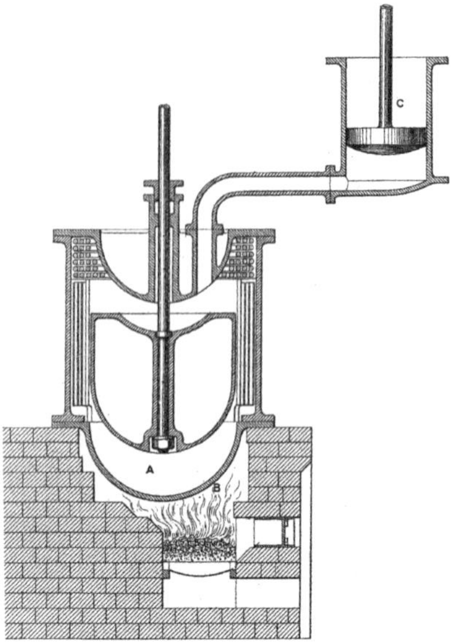

Fig. 2.5 Design of the displacer and power cylinders of a double-acting type of engine by Robert and James Stirling in 1827. After Jenkin (1885) with the permission of the Institution of Civil Engineers

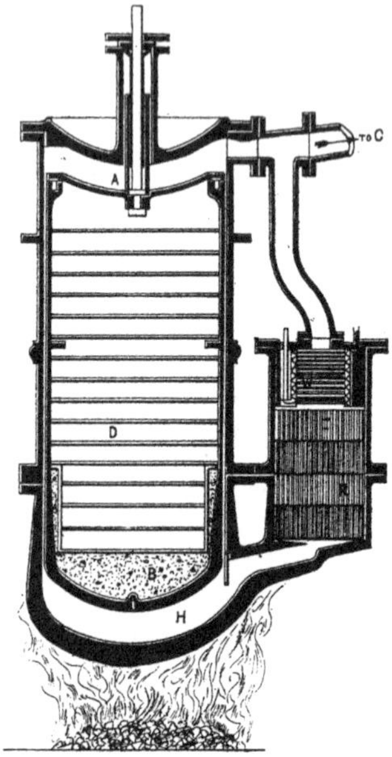

Fig. 2.6 Further improvements to the double-acting Stirling engine, patented in 1840, with regenerator and cooler in a duct outside the displacer cylinder. After Kennedy (1904–5)

the engine would work very well without cooling when starting up, the power tended to drop when the upper ends of the displacer cylinders became hot. This suggested the use of a separate cooler. The stationary regenerator allowed the provision of an efficient cooler at entry to the cold space, made of coils of copper pipes through which cold water circulated.

It is evident that with Stirling's earliest engine, shown in Fig. 2.1, air cooling by natural convection was quite sufficient to balance the heat input by the heat transfer from the flue gases at the upper end. With the improved models, where the cylinder bottom was exposed directly to the flames, this was no longer the case, and water cooling had to be used.

One disadvantage of these twin engines, when compared with the original single-cylinder design, was that the volume of the 'cushion air', or dead space, was relatively large, which diminished the compression ratio and hence the output. In this engine the total volume of clearances at the cold end was 1.2 times, and at the hot end 0.31 times the swept volume of the working air (Popplewell 1897). In a well-designed engine these ratios could well be 0.25 or less. The weight of these engines was also enormous; for a 3.7 h.p. engine, 1 ton per indicated horsepower was quoted so that the specific weight was probably more than 3 tons per brake horsepower (Donkin 1911)

From experience gained with this engine, still further improvements were introduced by 1840, as shown in Fig. 2.6 (Jenkin *op. cit.*, Stirling 1845). With relatively short displacer cylinders in earlier engines, the upper ends still tended to become hot and in this model the whole cylinder was therefore made much longer. The regenerator *R* and the cooler *W* were transferred to a separate enclosure outside the cylinder, with three connecting ducts to the bottom *H* of the displacer cylinder and one to the pipe *C* leading to the power cylinder. The regenerator *R* was made of stacks of iron sheets 0.025 in thick, separated by ridges with passages about 0.020 in wide. To reduce longitudinal heat conduction, the regenerator was subdivided into four stacks. The cooler *W* was made of a nest of copper pipes 0.125 in bore and 0.250 in outer diameter, with gaps of 0.050 in between the tube walls. To minimize dead volume, the space between the regenerator stacks and the cooler was filled with pieces of broken glass, which also acted as part of the regenerator matrix. The cylinder bottom, which in all early hot-air engines was liable to burn out, was thickened up appreciably and an air pump (not shown in the diagram) with a delivery pressure of about 150 lb/in^2 was used. Great care was taken to make this engine airtight under the high pressures used. Quoting from the specifications: '.... *a small strip or wire of lead is inserted at all the other joints to render them airtight, and the whole is strongly bolted together, and being calculated to support 1000 lb, is proved by water pressure from within 700 lb upon the square inch...*'.

The displacer *D* was made longer and was filled with brick dust *B* at the lower end. Horizontal thin iron plates supported the walls, which must have resulted in a great reduction of the heat loss transferred through the displacer from the hot space to the cold space, by introduction of insulating layers of air. As in the previous model, care is taken that not only the power piston, but also the packing of the piston rod works in a cold portion of the engine. In this later arrangement there is an ingenious contraption to eliminate leakage of air through the gland. The piston rod is surrounded by a long, cylindrical pipe filled with oil. A leather collar is clamped at the upper end to form a skirt round the piston rod, while the lower end permanently dips into the oil. The idea behind this early seal may have been clever, but it is easy to see now that it could hardly have withstood any pressure, even at the very low speeds used, although no records to this effect were left.

By that time the inventors had realized the harmful effect of unnecessary dead space, more so than many later engineers, as can be seen from the following further extract from the specifications.

'... as every portion of space within the air vessel, cylinder, plate box and passages takes in a quantity of air when the pressure is increased, and gives out a quantity when the pressure is diminished, so as to impair the impulse given to the piston, we carefully guard against all unnecessary space in the interior of the engine and in particular we make the passages as narrow as is consistent with the easy motion of the air ....'

One industrial application of this engine was a 45 h.p. installation in a foundry in Dundee which is shown in Fig. 2.7. This was an old beam engine which had been rebuilt to Stirling's specifications. Rated at 45 h.p., its bore and stroke were 1 ft 4 in and 4 ft respectively, and it ran at the sedate speed of 30 r/min. The minimum and maximum cycle pressures were 10 and 15 atm and the terminal air temperatures were 100° and 600 °F respectively. With an output of 37 b.h.p., the coal consumption was claimed to be 2.7 lb per brake horsepower-hour and 'sometimes' even 2.5 lb per brake horsepower-hour (Stirling 1845). Most probably these figures were somewhat optimistic, as this would correspond to a specific power of about 3.3 b.h.p. per cubic foot swept volume – a figure too good to be true.

The engine was first started up in March 1843, and seems to have run satisfactorily until a cylinder bottom burned out in December 1845. When replaced, the second cover lasted until May 1846, and a third until January 1847. At that stage, the owners seem to have lost patience and had the engine reconverted to what it had been originally, namely, a steam engine. It is a great pity that modern engines do not lend themselves so readily to radical changes of working principle – an obstreperous modern diesel engine could hardly be converted to run on steam, or vice versa.

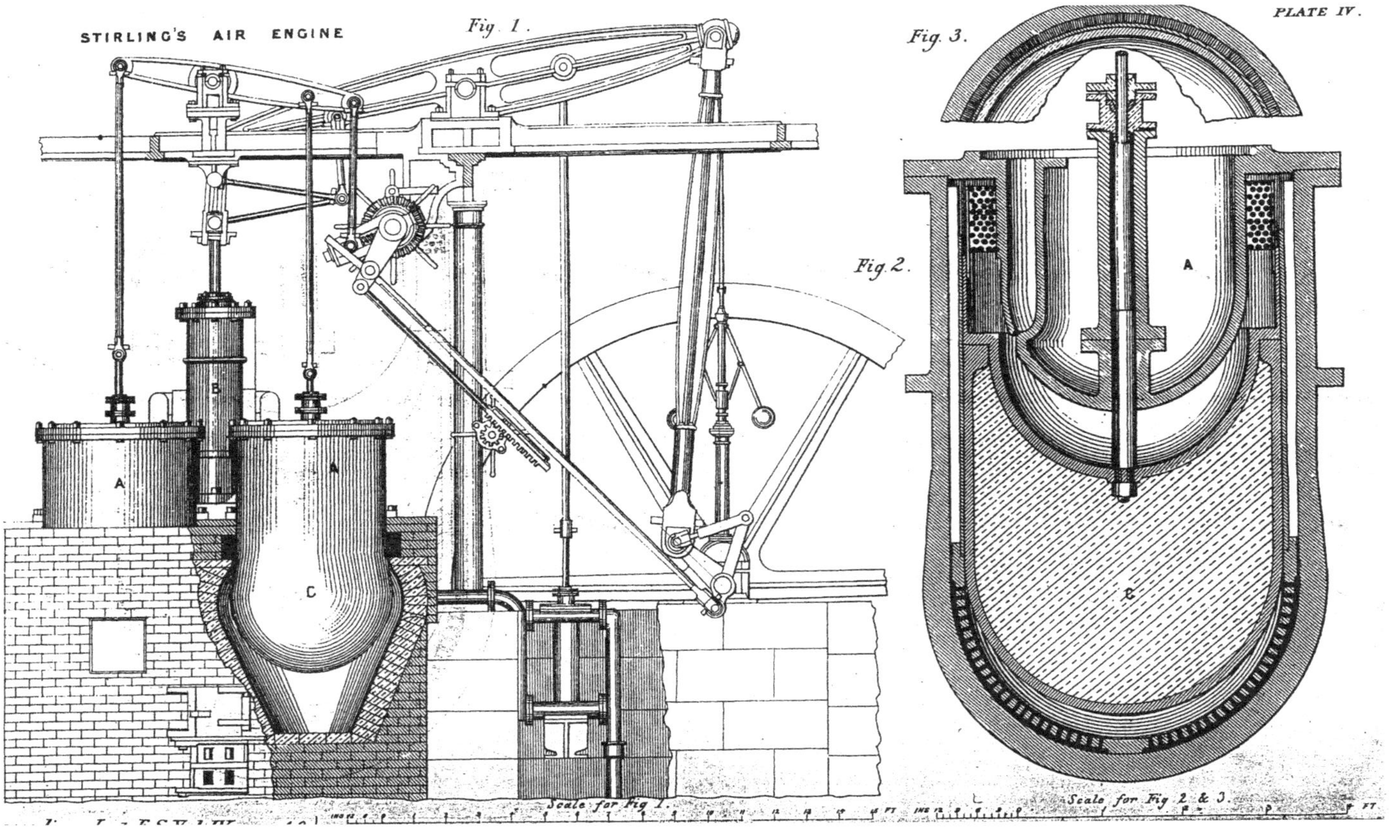

Fig. 2.7 Beam engine converted in 1843 into a two-cycle Stirling engine and rated at 45 h.p. at 30 r/min. Reproduced with permission of the Institution of Engineers and Shipbuilders in Scotland

It can now be appreciated that these engines failed to achieve a greater commercial success at that time chiefly because heat-resisting metals were not available for the cylinder bottoms. In steam engine boilers the surface temperature is always kept near the temperature of the boiling water, so that the metal is far less likely to overheat.

In air engines a delicate balance must exist between the heat supplied by the fire and that transferred to the working fluid if the metal temperature is to be kept within its creep strength.

Apart from the rather limited life of the cylinder bottoms, the performance of these Stirling engines seems to have been quite satisfactory in comparison with alternatives available up to 1850. Stirling (1845) claimed an economy in fuel in the ratio of 6:26 compared with steam engines. Although this was probably somewhat exaggerated in spite of Stirling's precise numerical formulation, it was generally conceded at the time that under favourable conditions Stirling engines were superior in output and efficiency to available steam engines, in spite of a mechanical efficiency as low as 39 per cent (*Anon.* 1875, Donkin 1911).

2.4 Degenerate Stirling engines – double-cylinder types

Stirling's working principle was used in many different engines built after 1860 in a number of countries. At first most of these used Stirling's second design of 1827, placing the power piston in a separate cylinder, but without making this piston double-acting as in the original prototype. This resulted in a single-cycle, two-piston, two-cylinder construction – and a somewhat lower mechanical efficiency than Stirling's own two-cycle, three-piston, three-cylinder design of 1827, and much lower efficiency and output than Stirling's first single-cycle, double-piston, single-cylinder construction of 1816. The main difference between most of these later engines was the position of the power piston in relation to the displacer piston and several alternatives were tried. For example, in the engines shown in Figs 2.8 – 2.10, the power cylinder is turned upside-down, in those shown in Figs 2.11 – 2.14 it is turned through 90°, and in Fig. 2.15 its alignment is the same as that of the displacer cylinder.

In Fig. 2.8 an early commercial application is shown, which was a compact and simple pumping plant comprising reciprocating water pump, a prime mover to drive it, and a furnace. This engine was made by Sir W. H. Bailey and Co. Ltd of Salford, Manchester, who manufactured different types of hot-air engines in substantial numbers between 1860 and 1908. From the cross-section shown in Fig. 2.9 it will be seen that the complete installation has only seven main moving parts. This extreme simplicity made the engine particularly suitable for use by unskilled labour, and it was claimed that no more attention

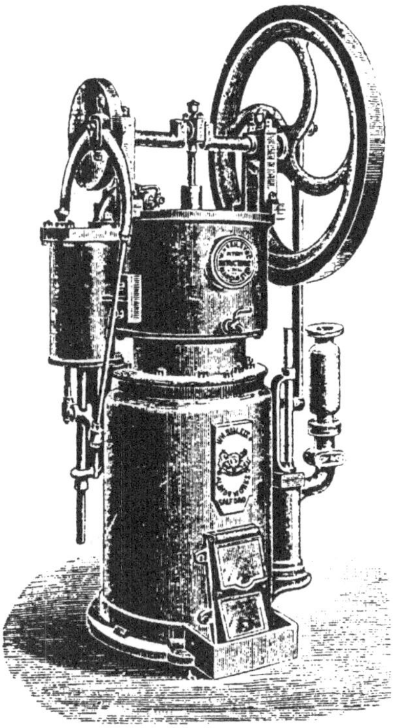

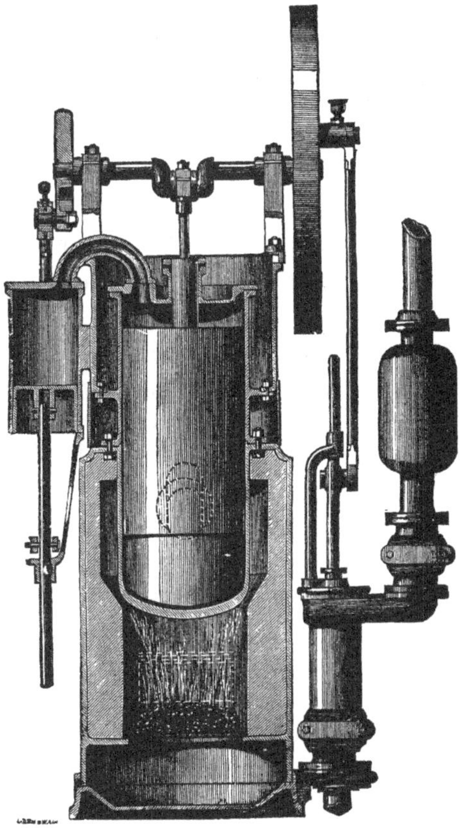

Fig. 2.8 Woodcut of a Bailey vertical hot air pumping engine with integral water pump. Built in substantial numbers between 1860 and 1900. (With permission of IMI Bailey Birkett Ltd)

Fig. 2.9 The cross-section of the engine shown in Fig. 2.8 (after Kennedy, 1904–5) shows the extreme simplicity that was achieved. This prime mover had only five main moving parts: crankshaft, two pistons and two connecting rods

or skill was required than for looking after a slow combustion stove. The link mechanism is considerably simplified, and sliding crosshead guides are used instead of straight-line motions. As in the two-cycle Stirling engines, the displacer moves in a cylinder directly above the furnace, with a passage leading to a separate power cylinder. The water pump is an integral part of this engine and the water is passed through the cooling jackets round the upper portion of the displacer cylinder. This engine is rather primitive and worked without regeneration or raised pressure level and its efficiency could not have been more than a small fraction of that of the original Stirling engine (Kennedy 1904–5). Figure 2.10 shows an early woodcut of such an engine on a farm in Guernsey, Channel Islands, pumping water for tomatoes in a large greenhouse.

Engines of this type were still included in a catalogue of Bailey's as late as 1893. Three different models were then produced with net power outputs of 1/16, 1/8, and 1/4 h.p. respectively. From the weights given, their specific power ranged from about 7.5 tons/b.h.p. for the smallest to 4.8 tons/b.h.p. for the biggest models, computed from the flow in gal/min and the head quoted in the manufacturer's pamphlet.

About that time the 'Robinson engine' appeared on the market, which differed from this early design only in using a horizontal cylinder instead of a vertical one. These engines were made by A. E. and H. Robinson and Co. and under licence by L. Gardner and Sons Ltd and Pearce and Co., all of Manchester, between 1895 and 1914. Many thousands of these engines were sold and they became popular for applications where a small and simple source of power was needed. Such engines could be driven by a small oil lamp or a gas burner instead of using solid fuel, and were frequently referred to in catalogues as 'Domestic Motors'. These engines also had separate displacer and power cylinders and, as was the case for the engine described previously, were single-cycle, two-cylinder, two-piston designs, but with the cylinders placed at right angles to each other. The last engines of this type were marketed about 1920 by Norris Henty and Gardners Ltd, of London and Manchester.

An early type of this engine is shown in Fig. 2.11. This was sold in two models, advertised as the 'one manpower motor' and the 'two manpower motor' respectively. The displacer *I* and the associated cylinder *A* are similar to earlier designs and here again no regenerator was used. A wide passage *D* leads to the horizontal power cylinder, with a piston driven through a link mechanism which results in a phase difference of 90° in the piston movements. A water-jacket was fitted at the upper end of the displacer cylinder. The description of this engine also mentions a 'safety device' which prevented the engine from racing on light load. This consisted of a light spring-loaded disc interposed in the passage between the two cylinders. Normally this caused negligible obstruction, but when the engine speed, and hence the air speed, rose

Fig. 2.10 Reproduction from a woodcut dated about 1890 showing the layout of a greenhouse for tomato growing in Guernsey (Channel Islands) with a vertical hot-air pumping engine for the water supply (Source unknown)

Fig. 2.11 Cross-section and external view of a small Robinson engine with a link driving mechanism advertised in 1895 as a 'one manpower motor'. After Kennedy (1904–5)

too much, this disc closed, until the speed had dropped sufficiently. With no regenerator or raised pressure level, these engines could hardly have produced more than about one-tenth of a horsepower at 150 r/min, and the use of this speed limiter or of the curious contraption shown on the left side of the upper cylinder in Fig. 2.11, which was probably a belt-driven governor, is surprising. The probable explanation is that in the absence of any provision for dynamic balancing, the heavy reciprocating parts might have caused the engines to bounce up and down or topple over at a few hundred revolutions per minute, as there was no visible provision for bolting them to a foundation.

The model shown in Fig. 2.12 is similar but incorporates a number of detail improvements, the most important being the use of a regenerator made of a mass of thin wire, completely filling the displacer body. The displacer drive is also simplified by using a roller and a slipper pad worked off the connecting rod as shown.

Figures taken from an old catalogue show that a typical engine about 1 ft 11 in high weighed 168 lb and produced 0.045 b.h.p. Larger engines of this type delivering up to 5/8 b.h.p. per cycle were also made later and are shown in Figs 2.13 and 2.14. In Fig. 2.13 an engine built in 1908 is shown driving an attached pump by means of a chain sprocket, and Fig. 2.14 shows an example of a two-cycle engine, made simply by duplicating the components on a single crankshaft. This last construction dates from 1915 and it is surprising to find that nearly 100 years after Stirling's invention, engines as primitive as that should still be marketed for industrial applications. Most of these engines were built for export to countries such as South America, Africa, and the Far East.

A very interesting engine of the two-cylinder Stirling type was developed by Laubereau and produced in Berlin by Schwartzkopff in about 1860. Initially these were horizontal engines with a crank mechanism (Delabar 1866, Schmidt 1861). In 1864 improved vertical machines with a cam displacer drive were produced as shown in Fig. 2.15. Again no regeneration was used and the mean pressure level was atmospheric. In addition to these disadvantages, the power cylinder was made to communicate with the hot end of the displacer instead of with the cool end, in order to reduce the dead space in the communicating duct, so that entry of the hot air below the power piston must therefore have increased the difficulties of lubrication. It will be seen from the illustrations that a novel feature of this machine was the use of a cam to actuate the displacer. Such an arrangement has something to recommend it on theoretical grounds, as it is possible to make provision for a dwell at dead-centre positions. Compression and expansion may thus be more effectively confined to low- and high-temperature spaces respectively. The cam used here imparted upward motion for 120° of crankshaft rotation, rest during 60°, downward motion during 120° and rest during the remaining 60°. The design of this cam is

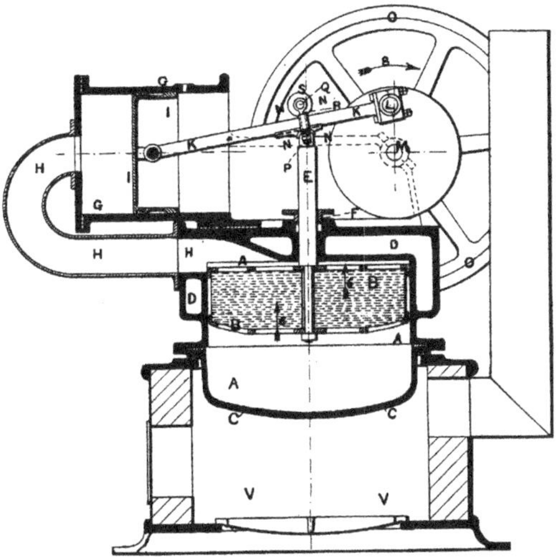

Fig. 2.12 Cross-section through an improved model of a Robinson engine which was fitted with regenerator and a roller drive for the displacer. After Kennedy (1904–5)

Fig. 2.13 Coal-fired Robinson engine, *ca* 1908, with attached chain-driven water pump. (Source unknown)

Fig. 2.14 Inefficient and badly designed two-cycle air engine, made in 1915 by placing two Robinson engines side by side. (Source unknown)

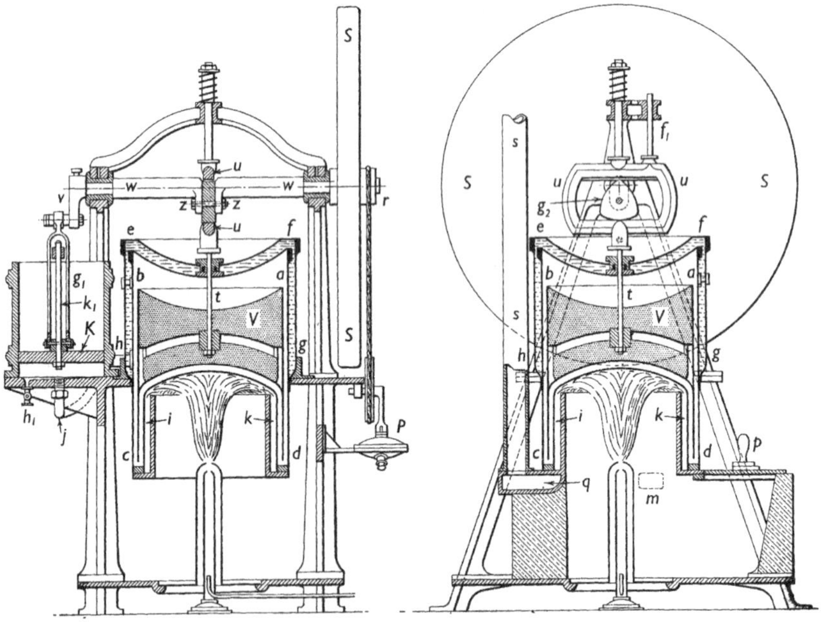

Fig. 2.15 Cross-section through the Laubroy–Schwartzkopff engine, which used a cam for imparting an intermittent drive to the displacer. (Source unknown)

explained by Weisbach (*op. cit.* at Article 475).

Another improvement in this engine is the use of extended surfaces for heat transfer. A cylindrical flame guide makes the hot gases flow along the surface of the cylinder bottom, and an extension shield is used with the displacer in a way similar to that used on Ericsson's engines, although the illustration is not very clear on this point.

The displacer cylinder *abcd* (Fig. 2.15) is surrounded by the water-jacket *egfh* at its upper end, which formed the cold space. In the lower end, the cylinder was closed by the bell-shaped cover. A hinged bottom cover *p* was provided so that the engine could be run either on solid fuel or by means of an oil or gas burner, as shown in Fig. 2.15. The hot products of combustion passed through the space *ik*, through the outlet *m* into *q* and out through the chimney *ss*. The transfer piston *V* consisted of two parts which were made of a material of poor thermal conductivity. The displacer cylinder was connected to the power cylinder by means of the pipe *j*. The power piston *K* had a piston rod k_1 working on vertical guides and was connected to the crank *v* on crankshaft *ww* by means of the forked connecting rod g_1. The driving arrangement of the displacer was rather more complicated. Its piston rod *t* passed through the cover of the cylinder through a stuffing box. It is seen from Fig. 2.15 that the crankshaft *ww* had two small cranks *zz* at the centre, between which a cam of approximately triangular shape g_2 was mounted. This cam moved the rectangular frame *uu*, which in turn was connected to the displacer rod *t* and the vertical guides f_1. Adjacent to the flywheel *SS*, a pulley *r* was fitted to the crankshaft *ww* which drove a small circulating pump *P* for the cooling water. A cock h_1 was provided by means of which the pressure could be released to stop the engine. It was claimed that the speed of this engine was above 500 r/min (*Anon.* 1867).

The 'Kyko' portable fan, which is shown in Fig. 2.16, used to be made in fairly large numbers until recently by the Model Engineering Company for export to tropical countries. It is a self-contained portable fan, about 4 ft tall, and will run for 12 h on 1 pint of paraffin. It achieved considerable popularity among missionaries, traders, planters, and well-to-do natives in outlying districts. This fan had the reputation of being virtually foolproof mechanically and ran for many years without attention. The very large blades give a fair air current, as the writer can testify from personal experience. Although made in England, this device is much more likely to be offered for sale in a Far Eastern bazaar than anywhere in the country of manufacture.

Many thousands of small engines of this type were also produced as toys. Some of these fascinating gadgets, which were often beautifully made can occasionally still be picked up in junk shops, usually in perfect working order. The only hot-air engine which was still being made in England at the time of writing (1959) is shown in Fig. 2.17. This is a small toy engine, made by Davies Charlton Ltd, for working Meccano or similar models. Overall dimensions are 7.5 × 4 × 2.5 in, including a small methylated spirit lamp, which will run the engine for about 20 min at 800 r/min on one filling.

Fig. 2.16 The 'Kyko' portable fan which became very popular in the Far East. Its height is nearly 4 ft and it is operated by a paraffin lamp. (Source unknown)

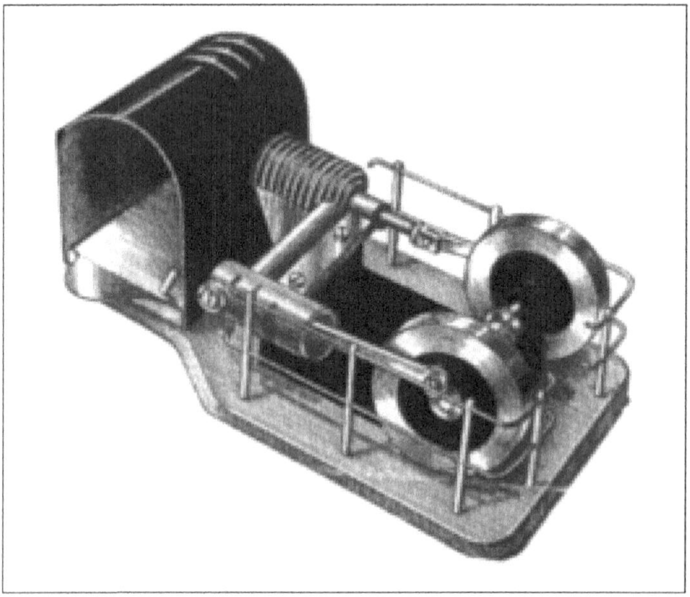

Fig. 2.17 Toy air engine in current production (1959) for driving small models, complete with methylated spirit lamp. Reproduced with the permission of Mrs A. Davies

3

Later single-cylinder engines

3.1 German air engines

It is evident that a closed hot-air engine with a power cylinder separate from the displacer cylinder works with a lower compression ratio than one with a single-cylinder design, due to the presence of a disproportionately large clearance volume. This considerably reduces the specific power, so that engines of the type described in the last section could hardly be suitable for applications other than mechanical toys or small models where efficiency did not matter. Although a patent registered in 1845 by Franchot described an engine with piston and displacer working in the same cylinder, Stirling's original design seemed to have been entirely forgotten for the next 50 years until the first engines by Lehmann were produced on the European continent. These engines were practically identical to Stirling's first design, except that they had a horizontal cylinder, direct contact between cylinder bottom and flames, and a more compact link mechanism to reduce overall size (Eckerth 1869a, Röntgen 1888).

An early model of the Lehmann engine is shown in Fig. 3.1. A very long cylinder *ABCD* was used, closed at the hot end and with the power piston *UU* reciprocating in the open end, which was kept cool by a water-jacket *LMNP*. About a third of the cylinder was enclosed by the furnace. The hot gases surrounded the lower end of the cylinder and kept it hot, after which they escaped through the smoke stack *RR*. The airtight hollow displacer was strengthened by a riveted plate *KK* to an extension of which the piston rod was attached. *EF* was a wooden plate closing the front of the displacer, and *HG* a domed metal cap closing the hot end. The weight of the displacer was taken on a support roller *O*. Between the displacer and the cylinder there was a narrow annular space through which the working fluid passed when transferred from one end of the cylinder to the other, but apparently no regenerator was fitted. The working piston *UU* was constructed similarly to a bicycle pump with a dished leather washer, which allowed additional air to enter whenever the internal pressure fell below atmospheric. The driving mechanism, by which the motion of the displacer was advanced by about 65° with reference to the working piston, was very similar to that employed on the Ericsson engine.

According to Eckerth such an engine with piston diameter 0.349 m and stroke 0.244 m, gave 0.984 h.p. at 97 r/min, with a thermal efficiency of 4 per cent, which corresponds to a specific power of 1.2 h.p. per cubic foot swept

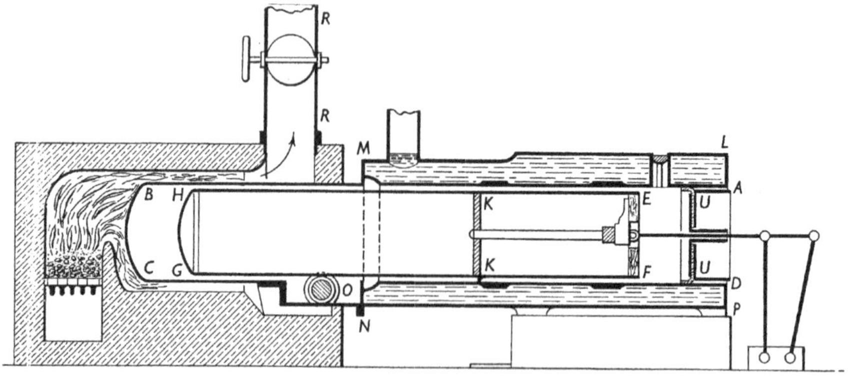

Fig. 3.1 Cross-section through an early Lehmann engine, made about 1860 on the Continent. The driving piston had a cup leather washer similar to that of a bicycle pump

volume. The mechanical efficiency was quoted as 65 per cent (Delabar 1864, Eckerth 1869b, Knoke 1899).

Lehmann's engine was immediately successful and large numbers were sold. Röntgen (1888) wrote enthusiastically:

'. . . (Lehmann's) engines are especially suited to the lesser industries. They are . . . more durable than Laubereau's and the annoying pounding of Ericsson's is entirely avoided. Moreover the consumption of fuel . . . is only half that for the other two systems. In this respect it is as economical as the best steam engines . . .'.

A number of different models to the same basic design were built later by Lehmann with horizontal or vertical cylinders. Improvements of detail were also introduced. For example, instead of the leather-cup-type of power piston a trunk piston was used, working in conjunction with a non-return valve called a 'Schnarchventil' (snoring valve) on account of its characteristic noise. In England this component was usually called the snifter valve.

Although these engines were operated without proper regeneration and without raised pressure level, they were fairly successful and were also manufactured in several other countries, initially under licence from Lehmann's original patents. For example, the woodcut in Fig. 3.2 and the cross-section in Fig. 3.3 show an engine manufactured by Bailey and Co. Again a very long displacer is used with rollers to take the weight, but also without a regenerator. The cylinder bottom is shielded from direct contact with the flames, which arrangement must have reduced the working temperature and hence also the power output, but which extended the life of the cylinder

Fig. 3.2 Horizontal 2 h.p. Lehmann engine, built in England during the latter part of the last century as the Bailey engine. After Kennedy (1904–5)

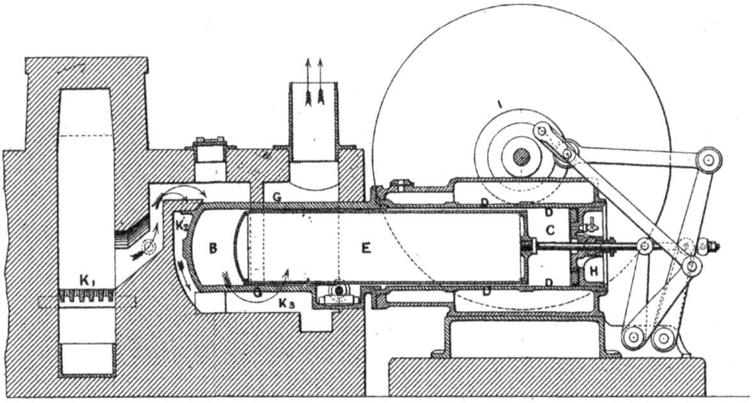

Fig. 3.3 Detail of the engine shown in the previous illustration. From Fleeming Jenkin (1885) with permission of the Institution of Civil Engineers

bottom. To save space, the two pistons were driven by sets of bell-cranks, with the crankshaft and flywheel set back along the cylinder. A separate water tank was usually fitted, connected by pipework to the large cooling-jacket at the open end of the cylinder. As in woodcuts of many other old air engines, a centrifugal governor is a prominent feature, although its function is probably more decorative than practical.

Figure 3.3 also shows a release cock fitted to the piston for speed regulation. Operating this control must have been a hazardous operation if the engine was running at any appreciable speed. It must have been a considerable feat of agility to get one's hand right in between the moving rods and levers and turn the cock just when for a brief instant it came near the cylinder opening.

This type of engine was made in six different sizes, with weights between 14 and 50 cwt, excluding brickwork, ranging from ¼ to 2 ½ nominal horsepower, and running at speeds of 120 down to 90 r/min. The specific weight of these engines had therefore been improved to between 1 and 2 tons/h.p., but was probably still worse than that of the original prototype by Stirling. It was widely advertised and sold in great numbers for driving pumps, churns, chaff-cutting, grinding, crushing and various 'Domestic Machinery'. It was also claimed that the brickwork could be erected in one day by an ordinary bricklayer and that the roomy fireplace held enough fuel to keep the engine running unattended for extended periods. Almost any kind of fuel could be used and its simplicity was a great sales feature. Quoting from a contemporary catalogue:

'The stove is provided with a roomy fireplace, holding enough fuel to keep the Engine going from 3 to 6 hours without stoking. The furnace will burn almost any fuel – coke, coal, peat, wood, sawdust, spent tan, riddled cinders, etc. and as it consumes its own smoke it needs no attention during the intervals of firing. Thus it will be seen that the attention required is so simple, and of such a light nature, that any intelligent lad, workman, labourer, gardener, or domestic servant may learn to work it in a few hours with the assistance of our printed instructions, and attend to it without material interference with their other occupation.'

3.2 Heinrici, Bailey, and other variants

A smaller hot air engine of this type, produced by Bailey and Co. in six different sizes, is shown in Figs 3.4–3.6. The design was similar to the previous model and the main application was also for pumping. The engine shown in Fig. 3.4 had an output of about 1/4 h.p. and was fitted with an integral water pump which could be used for wells up to 120 ft deep. Figure 3.5 shows

Fig. 3.4 Small horizontal air engine with integral water pump. The pumped water was passed through the cooling-jacket before discharge. Delivery 400 gal/h against 50 ft head. (With permission of IMI Bailey Birkett Limited)

Fig. 3.5 Medium-size horizontal air engine fitted with twin-barrel pump. Delivery 2500 gal/h against 50 ft head. (Source unknown)

a similar engine with 1/2 h.p. output driving a double-barrel pump, while in Fig. 3.6 an alternative layout with pump and power take-off is shown.

The first 'Solar Motor', of which records exist, was built in 1872 by John Ericsson and is shown in Fig. 3.7. Between 1872 and 1883 Ericsson built about nine experimental solar engines, some using steam cycles and some air cycles. The engine shown here seems to be a closed Stirling engine, with piston and displacer as first used by Stirling. A parabolic mirror was used to concentrate the sun's rays on the lower end of the cylinder (Ericsson 1876, *Anon.* 1899, *Appleton Cyclopaedia*).

Lehmann's design was 'improved' by Stenberg, of Helsingfors, in 1877, whose engines were marketed under the pretty name of 'Calorisca'. They were similar to Lehmann's original design, except that instead of a crank mechanism a 'Bogenschleife' [cam scroll, (Knoke 1899)] was used for driving the displacer. As in Laubereau's later engines, the object was to provide for periods of rest at the end of the displacer stroke, so that distinct phases of isothermal expansion or compression may take place. This was achieved by a roller working in a groove shaped like the segment of a circle. As in Laubereau's arrangement, it is doubtful whether this constituted a substantial improvement on account of the increased mechanical losses that arose. Experiments made with a Lehmann, as well as with a Stenberg, engine showed that according to the indicator diagrams Stenberg's engines did, in fact, achieve a higher compression ratio, and as both engines had atmospheric pressure as the lowest pressure, a higher maximum pressure was therefore reached in the latter. However, the diagram in the Stenberg engine was not so 'fat' as that in the Lehmann engine, and calculation gave the same mean effective pressure for each. With no gain even in indicated work, the net power output was probably decreased due to the lower mechanical efficiency.

When the internal combustion engine was developed, the larger types of air engines were soon rendered obsolete, and smaller models appeared on the market. For example, the engine shown in Figs 3.8 and 3.9 marked 'Heinrici Motor,' but otherwise of unknown origin, was fairly common at one time. Standing about 2 ft 6 in high and weighing about 80 lb, its output was 0.028 b.h.p., as measured in performance tests carried out by the writer. This corresponds to a specific weight of about 1.3 ton per brake horsepower and a specific power of 3.4 h.p. per cubic foot swept volume – figures which compare favourably with those of larger engines. A rather amusing refinement, which can be seen in illustration, is the provision of a brake shoe actuated by a thumbscrew and bearing against the rim of one of the flywheels, by means of which the power could be 'regulated'.

Comparing the Bailey engine in Fig. 3.3 with this Heinrici Motor, it will be seen that the link mechanism has been slightly simplified. The drive for the

Fig. 3.6 Horizontal dual-purpose engine installation for pumping water as well as supplying power. The gear driving the pump can be disengaged. (Source unknown)

Fig. 3.7 Sun motor designed by John Ericsson in 1872. Unlike most other engines constructed by him, this model appears to work on a closed cycle in accordance with Stirling's principle. After Ericsson (1876)

Fig. 3.8 Small 'Heinrici motor', late nineteenth century. Similar to Stirling's original engine, but without regenerator

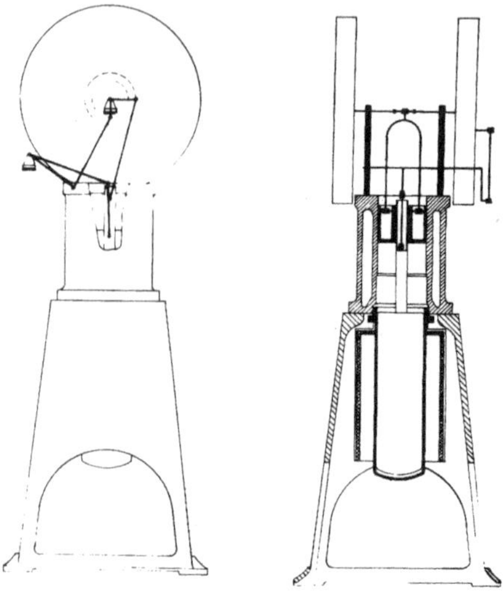

Fig. 3.9 Diagrams showing the internal construction of the 'Heinrici motor' illustrated in the previous figure

power piston is now more direct from a crank to a forked connecting rod, attached directly to the piston by two gudgeon pins. Only the displacer is still driven by means of a rocking lever and two connecting rods, which are actuated by a crank on the outside of one of the flywheels. This is shown clearly in the drawing, Fig. 3.9. A similar small engine from an illustration made by Gardner Ltd, and dated 1908, is shown in Fig. 3.10. Here the same crank is used for driving both the piston and the displacer, and the rocking lever is set high so as to produce the required phase difference between the motions of piston and displacer.

3.3 The Rider engine

It will have been noted that all the closed hot-air engines described so far hardly differed from either one or other of the two engines invented by Stirling,

Fig. 3.10 Air engine made by Gardner Ltd, in 1908 with redesigned link mechanism. Reproduced with permission of Messrs L. Gardner and Sons Ltd

and what differences there were often seemed to reduce rather than increase output or efficiency. In actual fact, right up to the middle of the present century, only one single major innovation was introduced, when in 1876 a new type of hot-air engine was invented in Philadelphia by A. K. Rider. The principle of this engine was that instead of varying the total volume of the working fluid by a single piston, and using a displacer to move the working fluid back and forth between the hot and cold space respectively, as has been explained with reference to Fig. 2.2, two separate pistons in two separate cylinders were used to achieve the same effect. Although the type of engine which was first produced in accordance with this principle was of no great practical significance, this innovation prepared the way for the novel multi-cycle air engines proposed by Rinia in 1947.

The sole manufacturing rights of Rider engines for England were acquired by Hayward Tyler and Co. Ltd, in 1877, and about 500 were sold within the first 5 years. Three sizes were made, nominally ¼, ½, and 1 h.p. respectively, and most of these engines were fitted with a water pump as an integral part. They were similar to the engine shown in Figs 3.11 and 3.12, where exterior views and a cross-section respectively are given.

Referring to Fig. 3.12, two loaded pistons *C* and *D* reciprocating in the cold cylinder *A* and the hot cylinder *B* respectively were used, and the cylinders were connected through the regenerator *H*, so that the working fluid traversed the regenerator whenever it passed from one cylinder to the other. The lower portion of cylinder *A* was surrounded by the cooler *E*, through which the cooling water circulated. The heater *F* was kept at a dull red heat by the fire below. Extended heat transfer surface was provided in the shape of a cylindrical skirt, so that the air was forced to pass through a narrow passage near the hot cylinder wall to facilitate the transfer of heat. A contemporary description ran:

'The heating and also the cooling of the air is instantaneously effected by its alternate presentation to the surfaces of the heater and cooler in a thin sheet, such being found by experience to be the only correct method of rapidly and thoroughly effecting changes of temperature in air.'

Two sets of leather packing rings *K* were used to make the two pistons leak-proof. In order to protect the piston and the packing in the hot space from the adverse action of the heat, the upper end of the hot cylinder *B* was kept cool by means of a water-jacket. In the engine shown the pumped water was also led through the cooling-jackets, passing first through the pump, then through the main cooling-jacket *E*, and lastly through the smaller jacket on top of the hot cylinder, before delivery from the flanged opening seen in front of the cylinder in Fig. 3.12. An automatic valve *L* at the lower end of the cold space was

Fig. 3.11 Reproduction from two woodcuts showing the external appearance of a Rider engine with integral water pump. Half the furnace can be swung forward to gain access to the hot cylinder. Reproduced with permission of Messrs Hayward Tyler Ltd

provided to admit air whenever the internal pressure fell below atmospheric. Cock *M* could be opened to atmosphere to stop the engine. The water pump was attached directly to the cold piston by a rigid connecting rod. As in many other hot-air engines, the cylinder bottom in the hot space used to burn out frequently due to the poor materials available at the time, but replacement was made particularly simple by swinging half the furnace forward, as shown in Fig. 3.11. In this engine the phase difference between cranks was made 95° and the regenerator consisted simply of 24 plates of cast iron with a distance of 2 mm between them (*Anon.* 1876)

Although Rider engines worked with a lower mechanical efficiency than Lehmann engines, the utilization of heat was better, as a regenerator was used. Calculations of the efficiency of this regenerator and experiments in which the engine was run with and without this regenerator, showed that such a regenerator had an efficiency of 60 per cent (Schöttler 1881). Schöttler was impressed by this result but, compared with values of well over 90 per cent attainable today, this is a very low efficiency. It is due partly to the relatively large dimensions of the flow passages – but there is another consideration described below.

The regenerator is sited in the wrong place! It sits at the *cold* end of the expansion plunger and at the *warm* end of the compression plunger. If the engine boasted a degree of thermal regeneration, much of it must have derived

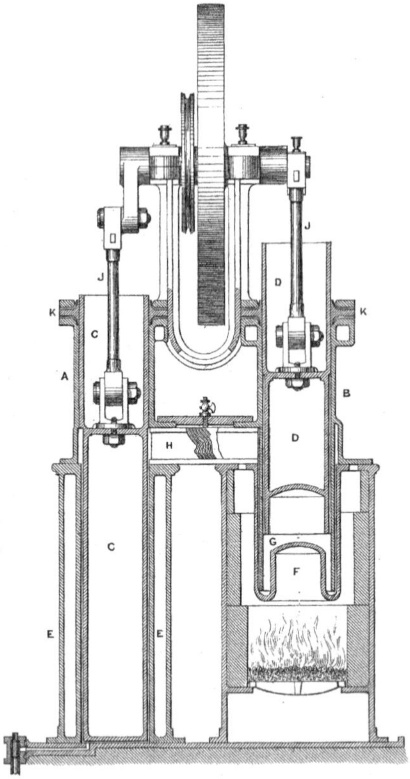

Fig. 3.12 Cross-section through the engine shown in the last illustration. The extreme simplicity – there are only five moving parts – should be noted. After Jenkin (1885) with permission of the Institution of Civil Engineers

from flow through the annular passages around the plungers. To this extent, the type did not completely presage the later, Rinia invention, which placed the regenerator appropriately between the high- and low-temperature parts of the engine.

However, the simplicity of the Rider and the ease of operation were greatly appreciated by the public. The advertising slogan used at the time by the makers was: *'These Engines work WITHOUT EXPLOSION'*, to distinguish them from the internal combustion engines which were then beginning to appear on the market. One letter from a satisfied customer in Portugal is reproduced below, as printed in an old catalogue:

'Gentlemen, the first engine you sent me (No. 2 Pumping Engine) is at work in a 'Quinta' near here. It does more than was expected from it, and with great economy. Four to six oxen were required to do the same work, and they did it

badly. A lad of 12 years is working it; he breaks up the wood, irrigates the orange grove near the Engine, and has time to spare. Yours truly, Thos. G. Hall, Lisbon, May 20th, 1878.'

About 90 per cent of all the Rider engines sold were used for pumping water. The rest were for miscellaneous applications, where moderate amounts of power were needed. Many church organs, for instance, were at that time motorized by a Rider engine fired by gas. This must have compared favourably in silence and reliability with the young boys that had been used previously for supplying the motive power, as witnessed by the following letter:

'Dear Sirs, If you procure a copy of the Musical Times for July, you will see how highly I am able to report on the Engine. I know of nothing (in all points) so perfect – no trouble at all. The gas burner answers admirably, requiring to be lit for 20 or 30 minutes before I want to use the organ, consuming about 25 feet per hour. I have confidence in expressing my belief that no small power is so easy to manage, so economical and – which is of great value – so noiseless as this. Yours truly, C. H. Fynes-Clinton, Blandford Rectory, Dorset, July 3, 1879.'

The total number of these engines that were sold between 1877 and 1895 was approximately 1000, including one to the Prince of Wales (later Edward VII) for Sandringham, and one to the Khedive of Egypt for Ras el Tin Palace in Alexandria. This latter engine may well have been ordered to replace an Ericsson engine imported from New York. They were also adopted by H. M. Government for water supply to ports and barracks, and records exist of engines that had been in continuous use for 20 years without major repairs. A description of an engine delivering 250 gal/h water against a head of 390 ft, burning 9 lb of coke per hour, can be found in *Engineering* for 1881.

The possibility of using solar energy for driving a hot air engine was also brought up again. A Rider engine was proposed as a solar power plant for Phoenix, Arizona, in 1908, as shown in Fig. 3.13. Although on first sight this scheme looks promising, an elementary calculation would show that the area of the solar absorber is far too small to deliver any appreciable power output (Yellott, 1957). Although the last Rider engines were produced in the mid-1920s, only very few engines had been made since about 1905.

The design of Rider engines and of other models remained static for a long time, although improvements in detail were put forward. For example, regulation of hot air engines by methods which are less wasteful than dissipation of energy through braking at the flywheel or escape of the working fluid were suggested by Slaby and put into practice by Buschbaum (Knoke 1899).

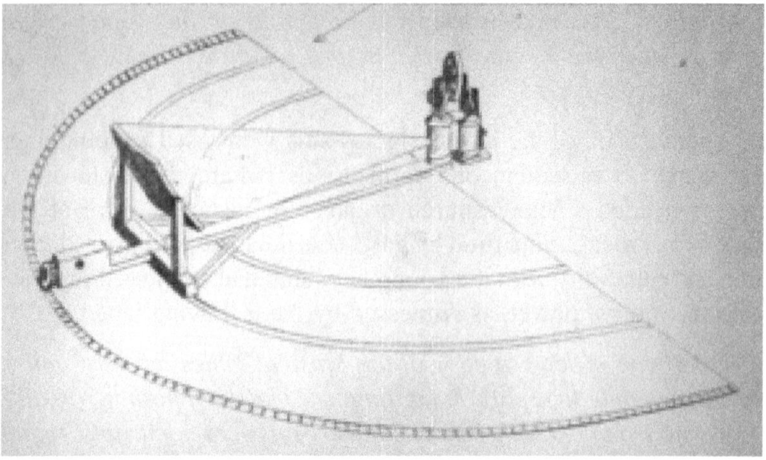

Fig. 3.13 Proposal for a solar power plant for Phoenix, Arizona, dated 1908. This was to use a Rider hot air engine but is evidently impractical on account of the small concentrator area

Some of the engines also reverted to Stirling's old idea of raised pressure level, and the provision of a 'cooler' and 'heater' became more general. These took the form of narrow passages whose walls were kept at the requisite temperature, and through which the working fluid had to pass whenever it was transferred from the hot space to the cold and vice versa.

4
Philips engines

4.1 The rediscovery

Between the two world wars little use was made of air engines other than for trivial applications, such as scientific toys, or for very small engines for special purposes, such as liquid stirrers in laboratories or powered fans for the Far East. Although many older engineers still affectionately remember small air engines as the first mechanical toy they ever played with as boys, the working principle was nearly forgotten, until even textbooks on thermodynamics included little more than a brief reference.

This subject was reopened a few years before the last world war at the Physical Research Laboratories at Eindhoven, then under the direction of Professor G. Holst. Tests showed that the overall thermal efficiency of air engines available at the time amounted to less than 1 per cent, compared with a figure indicated by the theory underlying the Carnot cycle, which may be up to 50 times as great. It is reported that this enormous difference between actual and theoretical performance caused Professor Holst to embark on this research, which was undertaken by a large team of scientists working under H. Rinia and continued partly in secret during the war under German occupation. Some leakage of information occurred, however, and towards the end of the war a German 'Spreng Kommando' was called in to remove a novel type of secret engine that was being developed. This they did, including some cylinders believed to contain a new secret fuel, which, on closer inspection, turned out to be nothing but compressed Dutch air (*Anon.* 1947).

In the course of this research work the fundamental properties of Stirling engines in general were investigated. For this purpose a number of different experimental models were built, some using the Rider construction with two loaded pistons, others the Stirling single-cylinder principle with one piston and one displacer. When the war ended, a number of successful small prototypes had been produced with quite startling performance compared with the old-type air engines (Rinia and duPré 1946). It was claimed that for equal powers the swept volume had been reduced by a factor of 125 and the weight by a factor of 50, comparison being made with a model still in production in 1923 (deBrey *et al.* 1947). Yet the general principle, as well as the layout of these single-cylinder engines hardly differed from Stirling's original engine of 1816, as can be seen from Fig. 4.1, which is a simplified section through the engine referred to. The photograph in Fig. 4.2 shows the external appearance.

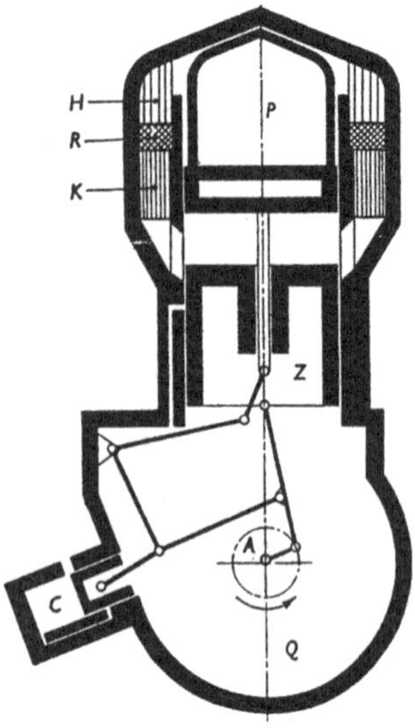

Fig. 4.1 Simplified cross-section through a Philips' single-cylinder engine of 1946. Comparison should be made with the original Stirling engine of 1816 shown in Fig. 2.1. Reproduced with the permission of Philips Electronics BV

Fig. 4.2 Exterior view of the single-cylinder Philips' 2 h.p. air engine. The gas burner has been removed for clarity. Reproduced with the permission of Philips Electronics BV

Comparing these with Fig. 2.1, it will be seen that apart from incorporating the driving mechanism in a crank-case, no fundamental changes have been made. Yet this modern engine is evidently a product of the twentieth century, and bears a striking resemblance to air compressors or single-cylinder motorcycle engines in current production (van Weenen 1947). This model combines the improvements and good points of all the earlier models with an application of modern theories in heat transfer, aerodynamics and strength of materials. In its construction use has been made of newly available materials, such as heat-resisting steel alloys. This modern-looking small engine has an output of about 2 h.p. and runs at a speed of up to 2000 r/min with a pressure range of 20–50 atm. It is claimed that the power output and efficiency are comparable with those of internal combustion engines, while a number of important advantages are obtained.

The cylinder top is heated by means of a gas flame, but in the photograph shown here the burner has been removed in order to show the appearance of the cylinder head. This dome is made of a thin shell of heat-resisting alloy to take the mechanical stress, and is provided with fins made of a good heat-conducting material, both externally and internally, in order to improve heat transfer. The power piston *Z* (Fig. 4.1) moves at the lower end of the cylinder, which is kept cool by means of a water jacket. The displacer *P*, attached to a piston rod, passing through *Z* is covered at the top by a thin cap of nickel–chromium steel. This reduces heat losses via the transfer piston and also protects the sliding areas from the direct action of the hot gases. *H*, *R*, and *K* are the heater, regenerator, and cooler respectively, arranged in an annular duct surrounding the cylinder. *H* is simply formed by fins on the inside of the outer cylinder wall, *K* is a series of tubes through a water jacket, and the regenerator is a matrix made of thin wire by a special manufacturing process. The greatest gain in efficiency was achieved by careful scientific design of these elements. It was necessary to minimize the size of the heat exchangers without reducing their performance, and without unduly increasing the flow resistance. At the same time, a very good regenerator design was achieved, and efficiencies over 95 per cent were claimed for this component.

The transmission links and the crankshaft *A* are enclosed in the crank-case *Q*, which is kept at an elevated pressure by means of the small air pump *C*. A small passage leads from the crankcase to a point uncovered by the power piston at the position of greatest volume, which roughly corresponds to the position of minimum pressure. The crank-case is thus kept constantly at a pressure corresponding to the minimum cycle pressure, and the only point where leakage can occur is where the motor shaft emerges from the crank-case.

It will have been noted that in spite of the enormous technological progress in nearly every other branch of engineering, in this instance there had been very little fundamental improvement to a design as proposed by its inventor. The only advance that was made up to that time had been Rider's scheme, using two loaded pistons instead of a single loaded piston and a displacer. This prepared the ground for a further development, i.e. the proposal of multi-cylinder, multi-cycle engines, where several cycles according to the Rider system may take place on opposite sides of double-acting pistons. Due to a reduction in the number of main moving components, there would be an increase in mechanical efficiency as well as other incidental advantages.

4.2 Double-acting types

This proposition was made by van Weenen (Rinia 1946) and was utilized in other engines constructed by the Philips Company at Eindhoven. Though certainly novel in its conception, a similar idea had been proposed much earlier by William Edward Newton in a patent dated 4 April 1853. This was for multi-cylinder air engines with alternate hot and cold cylinders, with individual cycles taking place partly beneath a piston in a hot cylinder, and above a piston in a cold cylinder. There is no record, however, that these proposals had ever been used in a practical engine (Bourne 1878).

Figure 4.3 shows the only model constructed according to van Weenen's proposals, of which details were published. It appears that this four-cylinder engine has only nine main moving parts, as far as can be told from what meagre information is available. Four double-acting pistons are used, working in four parallel cylinders in 'square' formation. The engine is particularly compact because a wobble-plate is employed here instead of a more conventional transmission with two crankshafts. Many features developed for the single-cycle model described above have been applied to this engine, such as annular heat exchangers and regenerators and the special design of cylinder heads and piston caps. This engine can be run in both directions and can be reversed by means of a slider while the engine is running, interchanging the connections from the heat exchangers to the cylinders. This engine is rated at 20 h.p. and runs at 3000 r/min.

According to information originally published, it was intended to proceed with the commercial development of hot-air engines as prime movers, covering a range from fractional horsepowers up to several hundred horsepower. It would appear, however, that in the course of this development the property of thermodynamic reversibility opened up new and even more promising prospects.

An engine constructed in accordance with Stirling's principle, as explained with reference to Fig. 2.2, is perhaps the only practical example of an ideal heat

Fig. 4.3 A 20 h.p., four-cylinder, double-acting Philips' air engine with wobble-plate drive. Reproduced with the permission of Philips Electronics BV

engine as first described by Carnot (1824). The fundamental property of such an engine is that when it is supplied with heat energy it will run as a prime mover and convert heat into mechanical power. Conversely it is also possible to supply mechanical power to drive such an engine, and it will then act as a heat pump and utilize this work in elevating heat to a higher temperature level. One early application of this principle was by Kirk in 1862, whose reversed air engine was reported to have worked for 10 years (Kirk 1873).

It was found possible to use the Philips' single-cylinder engine virtually without major modifications as a heat pump, achieving temperatures well below –200 °C, and to liquefy air without pre-compression. Thus a very simple and compact air liquefying machine had been obtained, and such units are being marketed and have been in regular use in a number of laboratories for some time. The appearance of an early experimental installation built in 1954 is shown in Fig. 4.4. The operator is seen drawing off liquid air, which has condensed on fins attached to the cylinder head inside the insulating cap seen on top of the machine.

This machine is, in fact, a remarkable innovation and offers advantages that could not be achieved by any of the conventional methods for liquefying air. It is far simpler in construction, and as the air is not subjected to pre-compression before liquefaction, an oil-free product is obtained. Delivery of liquid air is stated to commence within 15 min of starting up, compared with 2 h often required with conventional equipment. This development has been fully described elsewhere, and for a more detailed description the interested reader is referred to the accounts by Köhler and Jonkers (1954a,b,c) and Köhler and van der Steer (1955).

Fig. 4.4 Single-cylinder, two-piston gas refrigerating machine built by the Philips Company in 1954 for liquefying air at atmosphere pressure. Reproduced with the permission of Philips Electronics BV

4.3 Future possibilities

In the preceding chapters a variety of different types of old-fashioned engines have been described, all of which became completely obsolete and survived only as museum pieces. For a long time they were only of interest to students of engineering history and, to a more limited extent, to theoretical thermodynamics, in providing a hypothetical cycle of reference for other processes, while practical engineers made no use whatever of this principle. The preceding sections then showed how, in the course of the ambitious research programme by the Philips Company, machines with impressive performance were developed.

Full details of extremely efficient prime movers have been published more than 10 years ago, yet at the time of writing it still seems that no air engines are being built or used for purposes other than research and development. In the reversed form, however, Philips' gas refrigerators seem to have had considerable success as air liquefiers and are being used fairly widely for this purpose. The questions that spring to mind are the following:

(1) What will be the future of air engines in general?

(2) Does the long and varied development of air engines as prime movers represent no more than a dead branch on the evolutionary tree of technology, or will modern types of power producers soon come into general use?

(3) Will refrigerators with closed air cycles be employed for other purposes than liquefying air as well, and will they differ greatly from existing designs?

An attempt to answer these questions is made in the concluding section of this chapter.

Based merely on technical considerations, the case for a revival of this scientific principle and its use for a variety of purposes is fairly clear. Stirling engines are probably the only practical example of heat engines with a theoretical efficiency or coefficient of performance equal to Carnot's law, which is the highest that can possibly be attained. As a corollary, it follows from fundamental thermodynamic principles that such engines must be fully reversible in a thermodynamic sense. This is, in fact, true of Stirling engines, which can work either as prime movers or as heat pumps without internal modifications. In practice there are certain technical difficulties in approximating to this theoretical performance but, as published results have shown, these can be overcome by an application of modern knowledge in thermodynamics, heat transfer, aerodynamics, and theory of machines. Engines with considerable development potential can therefore now be designed with an inherent basic efficiency higher than existing types. There is, therefore, a very strong argument for devoting further attention to basic research into the fundamental properties of such systems.

The commercial prospects of air engines are rather more difficult to assess. It is evidently true that a radically different piece of machinery will be much more readily accepted for applications where no alternative equipment with a roughly similar performance is available. Where conventional types of machines can be obtained whose performance is adequate and which have already gone through a lengthy process of technical development, it takes a long time to establish a revolutionary new type of product in direct competition. This will be the case, even if there are certain marginal advantages in performance or in constructional features and, even more so, if the new product is somewhat more expensive, which at least initially is often likely to be the case. Practical application of Stirling engines in the near future will, therefore, probably be in fields where no alternatives exist which have corresponding performance.

The Philips' gas refrigerator is, in fact, a demonstration of this rule. It so happens that until recently there was a wide gap in the range of temperatures that could usefully be achieved with available refrigerating systems, and no practical principle that could produce temperatures between those obtained by evaporative types of refrigerators and by conventional air liquefiers was known. Wasteful methods, such as using liquid air or solid carbon dioxide, had to be resorted to for cooling in the range of, say, –80 to –180 °C. There was, in fact, a 'natural' opening for any new type of refrigerator operating within these limits, and it is quite likely that Stirling cycle cooling machines will continue to be used for this purpose. In view of the special advantages of the Stirling gas refrigerator, it would appear that the useful temperature range could be extended, possibly through the use of a radically different design. It follows, therefore, that there should be quite a number of different applications as heat pumps that can be foreseen at this stage. In theory, possible applications range all the way from cooling machines allowing experimental physicists to reach very low temperature levels, to heat pumps for elevating low-grade thermal energy.

As an isolated example of many possible constructions that could be used, an experimental prototype of a two-cycle, double-acting Stirling engine, designed and built by the writer for domestic refrigeration, is shown in Fig. 4.5. Running tests proved that the coefficient of performance was superior to that of conventional equipment. The advantages of equipment that uses an unpressurized circulating system without the use of a volatile refrigerant are obvious. In this way it is possible to construct kitchen cabinets with actuating units mounted remotely, for example, on the exterior of the house wall. Nevertheless, such a scheme would be in direct competition with vapour-compression and absorption-type refrigerators in common use today, and its ultimate success would depend largely on a variety of other factors, such as manufacturing costs, consumer preference, development expenditure, etc.

Before examining in detail other different future possibilities, it is well worth emphasizing that due to the prolonged stagnation of engineering development in this field, a great number of fundamental technical aspects remain to be investigated. These include such basic problems as cylinder arrangement, thermodynamic analysis, heat exchanger layout, operational parameters, etc. and it is intended to deal with many of these problems in a later publication. It is also pointed out here that it is always difficult to try to prophesy future developments, and the following notes are no more than a collection of random possibilities that might be forecast at this stage.

For applications in the moderate temperature range, as, for example, in air-conditioning equipment, mechanical efficiency becomes important. It is possible that for this purpose the usual piston–cylinder assembly might be substituted by sealed bellows units. Extremely simple and compact air conditioners could be developed in this way, which would give either a cooling or a heating output at the flick of a switch. Another possibility worth examining for applications where a blowing or ventilating effect, coupled with a temperature change, is called for, is the use of an open cycle somewhere intermediate between Stirling's and Ericsson's systems. Unlike the other two

Fig. 4.5 Experimental prototype of two-cylinder, double-acting Stirling engine for domestic refrigeration, built in 1957. Reproduced with permission of GEC Alstom Ltd

main cooling systems in use today, the physical size of Stirling-type refrigerators can be made much smaller. It is therefore possible to construct very small refrigerating devices, such as, for example, a 'cocktail cooler', which one could insert into individual glasses to cool their contents.

In the field of prime movers, it will again be appreciated that although it might be quite feasible to construct Stirling engines in the medium power range, say, up to 10 h.p., which could compete on level terms with internal combustion engines in efficiency, compactness and cheapness, they will only be successful if some special advantage can be shown. For example, for underdeveloped countries, where neither manufactured fuel, nor operators of even moderate mechanical skill may be available, the simplicity of such engines and their ability to run on any 'natural' fuel such as brushwood or even camel dung, might be the deciding factor. With a relatively moderate amount of expenditure on development, an engine could certainly be evolved which would be of great benefit to power-starved countries such as India or China.

Another possibility lies in building Stirling engines which contain dynamo armatures as integral parts in a hermetically sealed and pressurized crank-case, resulting in a durable, compact and efficient electrical generator from thermal energy for moderate powers in isolated locations, such as for lighting, working wireless sets or charging batteries. In this connection it may be said that the old scheme of using solar energy for actuating a hot-air engine is certainly due for a reappraisal, and experiments have, in fact, been carried out for this in recent years by Khanna in India (Yellott, 1957).

Still other applications could be based on some specific advantages Stirling engines have, such as silence, reliability or durability, and some examples of these are given here, selected at random. Small boats could be driven by Stirling engines either as outboard or inboard motors, which would be nearly inaudible in comparison with existing engines.

For use in atomic piles or with other thermonuclear devices, continuous running, inaccessible and self-contained prime movers could be built. For the liquefaction of gases for export overseas and the subsequent recovery of the thermal energy, reversible cycle engines may be called for. Lastly, a durable, simple and robust starter for gas turbines could be made.

An attempt has been made in this last section to highlight a few of the many possible future applications of the Stirling principle in a very wide field. It is worth repeating, however, that all the examples suggested here are no more than conjectures, and it remains to be seen which course this development will take in the near future. It is to be hoped that these observations may at least have stimulated further interest in a promising and fascinating subject that has been neglected for far too long.

4.4 Acknowledgements to the original four articles

The kind assistance of the following firms, who supplied illustrations, pamphlets or data from their records, is herewith acknowledged: Sir W. H. Bailey and Co. Ltd, Hayward Tyler and Co. Ltd, L. Gardner and Sons Ltd, The Model Engineering Company Ltd, and Davies Charlton Ltd. Thanks are also due to the Director, Science Museum, for permission to reproduce Fig. 1.4, which is Crown Copyright, Science Museum, London, to the English Electric Company (now GEC Alstom Ltd, Ed.) for permission to publish Fig. 4.5.

The authors are indebted to *The Engineer* magazine for permission to republish Chapters 1–4 which first appeared in *The Engineer* in 1959.

5

'Modern knowledge' . . . and all that

5.1 Now, where were we?

At the drafting stage, two options presented themselves for following on where Dr Finkelstein left off. The first was to attempt a seamless transition. This would have involved backtracking to somewhere in the mid-course of Philips' work and blending in the extra detail now available from sources such as Hargreaves (1991). In this case, there would, of course, have been no place for editorial comment. The alternative, based on the fact that the original account is complete in its own right as well as being something of a legend, was to leave it intact and to find an alternative way of bridging the gap.

The way the decision went will be self-evident, but the reason may not be. The Stirling engine unquestionably owes its renaissance of the 1930s to the Philips Research Laboratories. Early post-war accounts (e.g. that of de Brey *et al.* 1947) report improvements in power per swept volume in the ratio of 125:1, and in specific power of 50:1, attributing the gains to '. . *modern knowledge about heat transfer, flow resistance etc.*'. In context, 'modern' of course means contemporary with the Second World War. On the other hand, specific instances of the resulting insights are withheld.

If there is to be an objective assessment of the Stirling engine and its prospects, this must critically relate performance achieved to the technology available for design. Lacking specific information from the horse's mouth, the only course open is to construct an independent chronology of pertinent scientific steps. This fixes the start of modern developments considerably earlier than Philips' involvement, and at the point where the cycle was first the subject of scientific analysis.

The picture which emerges is in stark contrast to the everyday perception of the evolutionary course of scientific progress: a period of intense interest in the theory of air engines between the mid-1850s and about 1880 yielded scientific confirmation of the paramount role of the regenerator. A picture of the cyclic temperature excursions of the matrix was proposed, and it was appreciated that insight was to be had by following the thermodynamic processes undergone by a fluid particle as it travelled between the variable-volume spaces. Indicator (or *p–V*) diagrams and indicated cycle work were routinely calculated in terms of the precise volume variations of the particular engine under study.

The impact on the design of industrial air engines was, as the first four chapters have amply demonstrated, nil! For some reason unknown, there followed a half-century during which the scientific study of regenerative air cycles appears to have lapsed completely. The fact permits this account to be conveniently divided between pre-'Dark Ages' (1885–1935) and post-'Dark Ages'.

5.2 Pre-Dark Ages

Prior to 1854 the case for the new 'dynamical' theory of heat had been debated against the caloric theory largely as a matter of personal persuasion as Chapter 1 has illustrated. Rankine's treatise of 1854a changed all that. The imperatives of the Second Law of thermodynamics and of the mechanical equivalent of heat were argued graphically in terms of *process paths* – constant temperature (isothermal) and 'isodiabatic' lines drawn on the indicator (p–V) diagram. The impact was heightened by the fact that Rankine's study took the regenerative air engine as the vehicle for the exposition.

One of the examples was the closed-cycle air engine, illustrated on the indicator diagram by two isothermal curves linked by 'isodiabatic' lines of constant volume. Subsequent casual interpretation may be the origin of the 'ideal Stirling cycle' which has done such disservice to this field of study. Rankine was under no illusions: he modified his indicator diagram for the effect of the 'cushion' of air which did not go through the cycle. Later, in the context of discussion at the Institution of Engineers of Scotland, he pointed out that the 'ideal' diagram would call for the engine to function by an impossible sequence of jolts. On the other hand, it is perhaps a weakness of Rankine's approach that the regenerative process is treated as an abstraction with no regard for the practicalities of the necessary heat exchange processes.

It is perhaps surprising that Rankine's radical approach should be shortly followed by another. Without reference to any precedent, and evidently much impressed by the '*extreme elegance of the contrivance*' (of Messrs R. and J. Stirling), Lawrie (1860), undertook to short-circuit the '*costly and harassing*' processes of experimental development by offering a cycle analysis. By applying the ideal gas equation to the variable-volume spaces he expressed pressure as a function of crank angle, and from this derived an expression for ideal, indicated cycle work some 11 years before Schmidt's better-known exposition of 1871. If Schmidt's earlier paper of 1861a accurately reflects his own thinking at the time, he saw the (closed) cycle of the Laubroy–Schwartzkopf engine in terms of isothermal process paths joined by lines of constant *pressure* (an understandable inference from a 'real' indicator diagram taken from a machine with continuous piston motion) and, like Rankine, calculated in terms of 'process paths' – $T_4 = T_1(p_4/p_1)^{(\gamma - 1)/\gamma}$ and the like.

Carnot's overwhelming influence encourages a tendency to think of him as the sole voice from the other side of the Channel. Not so: Briot's account of 1869 usefully debunked the notion of the Carnot cycle as the unique 'ideal' engine, substituting practical regenerative cycles of the same limiting thermal efficiency. If we defer detailed reference to Schmidt's influential contribution of 1871, we find in 1874 another Frenchman, Hirsch, concerned about the '*ever-growing consumption of oil*'. (First oil well? Pennsylvania 1859.) Hirsch may have been the first to depict the closed-cycle air engine in opposed-piston form (Fig. 5.1a). He advanced beyond the notion of the regenerator as a mere thermodynamic concept, treating it as a physical entity calling for a small ratio of wetted volume to wetted area so that the temperature of a matrix element was essentially equal at any instant and location to that of the adjacent air. His resulting prediction of the cyclic temperature envelope (Fig. 5.1b) derives from the ingenious concept of the 'thermal wave' (*onde thermique*). The result is not strictly correct, but marks an important starting point.

Hirsch introduces 'Hirn's apparatus' (Fig. 5.1c.), which may be thought of as a Stirling engine with opposed pistons moving in synchronism. With total volume constant there is an increase in pressure as the air is displaced from the cold space to the hot. If the temperature of an air particle emerging on the hot side is the nominal exit temperature of the matrix, it subsequently *exceeds* this value due to the pressure increase accompanying the transfer. The point comes as a revelation even today, as does his tentative treatment of the distortion of his *onde thermique* by an inhomogeneous matrix.

Hirsch appreciates the insights to be had by tracking a fluid particle (a molecule) over a complete cycle of operations. It therefore comes as no surprise that his analysis provides for different parts of his *onde thermique* to propagate at different speeds due to the variation of density with location. He correctly restricts his predictions to cases where pressure drop and pressure wave effects can be neglected. The formulation of a Schmidt-type (mass balance) cycle analysis is, unremarkably, taken in his stride.

Against this background the better-known contributions of Slaby (1878) and Schöttler (1881) indicate a period of consolidation rather than discovery. Slaby's graphical solution of the equations for indicated cycle work appears to owe a good deal to that of Hirsch, and is interesting for coping with any desired piston motion. Schöttler essentially applies the Slaby approach to a different engine – a Rider. In the process he constructs a fluid particle path, but, unlike Hirsch, exploits it no further. (An exact path would reflect his temperature step at the regenerator.) Schröter's 1882 approach to regenerator analysis is in the Rankine mould, i.e. it is process-path based. It goes a step beyond in admitting real matrix materials such as woven wire mesh, and uses the Hirsch assumption of a common local temperature for matrix and fluid. Otherwise this is a

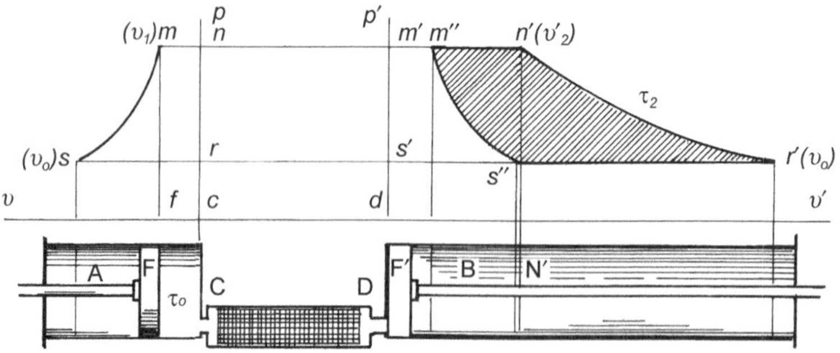

(a) A pioneering opposed-piston representation of the closed-cycle, regenerative air engine

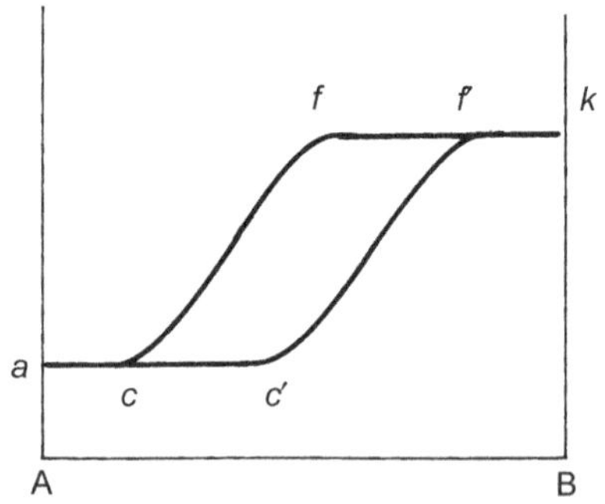

(b) Temperature envelope of regenerator in cyclic operation predicted from Hirsch's 'thermal wave' (*onde thermique*) approach

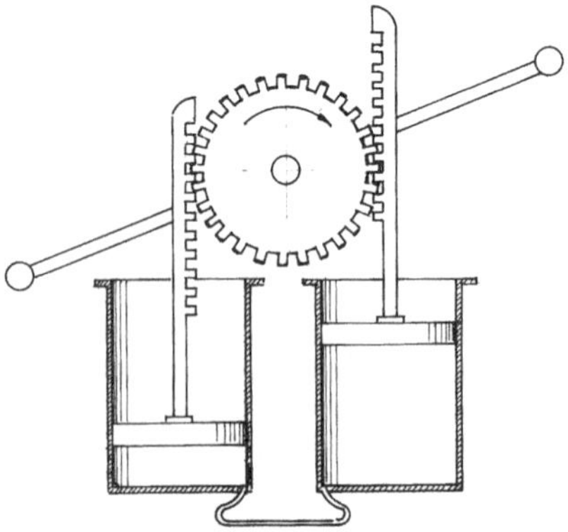

(c) Hirsch's depiction of Hirn's apparatus for demonstrating that a regenerator joining two vessels at temperatures T_E and T_C can lead to air temperature rising above T_E and falling below T_C

Fig. 5.1 Specimen diagrams from '*Theorie des Machines Aerothermiques*' by M. J. Hirsch (1873)

somewhat academic application of Second Law principles with little on offer for the engine designer.

Zeuner (1907) overcompensates for his earlier disparagement with a retraction and a deal of algebraic energy expended in deducing the 'mean' temperature in the regenerator. The objective – an estimate of the mass of air contained in the matrix – is worthy, but the starting point is the empirical Hirsch temperature distribution, and the result therefore arbitrary. Thereafter, the trail of scientific interest scrutiny in the regenerator runs cold. When interest begins afresh in the 1920s – this time with a serious analytical onslaught (Jakob 1957) – the context is not the air engine but the pre-heater for furnace air. There is no record of any impact on the air engine until the 1950s. Even then it is indirect: Schumann's elegant solution (1929) to the initial blow for the first time provides a means of obtaining heat transfer data for regenerator materials.

5.3 End of the Dark Ages

The focus of Schmidt's study in 1871 had been a specific air engine – the Lehmann engine, already illustrated in Chapter 3 – and yielded a prediction of indicated work per cycle and of the cyclic variation of pressure. Within the limitations of its starting assumptions it was equally applicable to all engines deriving from Stirling's original.

The analytical rationalization was delightfully simple: the equation of the ideal gas (Section 5.8.1 below) was applied to each separately identifiable volume of the engine – fixed, as in the regenerator, or variable, as in the compression and expansion spaces. To this extent it was no advance on Lawrie (1860). Only the fact that the 'Schmidt analysis' has taken hold to cover all treatments presupposing a fixed temperature distribution justifies elaboration here. For each of the individual volumes, the ideal gas law relates mass content to enclosing volume via temperature and instantaneous pressure. Algebraically equating the sum of the various fluid masses at their respective temperatures to total, invariant mass, M, permits the common factor, pressure, to be extracted and expressed as a function of crank angle. Integrating over a cycle with respect to total, instantaneous volume yielded the estimate of indicated cycle work.

It is *identically* this material – somewhat streamlined and obliquely acknowledged – which comes across as the crucial scientific tool in Rinia and duPré's 1946 paper on the resurrection of the air engine. Now, paraphrasing Paul Chambers of ICI, an organization such as Philips is in business to make money first and foremost, and to make electronic components, Stirling engines, etc. second. It would be bizarre if their scientific reporting did not reflect this.

Thus we probably have an inaccurate picture of their scientific priorities. However, on the available evidence, the *'modern knowledge about heat transfer, flow resistance, etc.'* appears to owe more to the period before the Dark Ages than to the twentieth century; Schmidt's treatment does *not* comprehend the regenerator, rather it supposes an arbitrary temperature profile. It is innocent of pumping losses, and so suggests that zero heat exchanger free flow area gives maximum indicated cycle work. Intelligently used it *may* indicate regenerator thermal loading, but as a tool for gaining insight into engine operation it is arguably inferior to Hirsch's approach of 1873. The potential for high thermal efficiency lies with the *actual* temperature response and corresponding flow-loss characteristics of the regenerator. On this basis, the true turning point in modern scientific understanding is the first analytical scrutiny of the regenerator.

5.4 The 'regenerator problem'

The clock must be turned back to the 1920s, and to a pioneering analytical study by Hausen (1927). Helmuth Hausen went on to devote a career lifetime to the study of heat exchangers. While a professor at the Technische Hochschule, Hanover, he published his substantial contribution in a book (1950), of which close to 200 pages dealt with the regenerator. His interest was in the application of regenerators to furnace air pre-heating. There is no shortage of reasons: furnace air pre-heaters are massive pieces of capital equipment, and are not cheap; accurate prediction of thermal characteristics promises economy of construction, improved temperature recovery, or both.

Before we go any further, a regenerator is a regenerator. From the analyst's viewpoint, the furnace air pre-heater, the Ritz rotary regenerator (for the gas turbine) and the regenerator for the Stirling engine or cooler are distinguished merely by the respective operating conditions. Variants for the first two applications have relatively high *thermal capacity ratio*. Later this will be given the symbol N_{TCR} and be formally defined. For now, a large N_{TCR} means that the matrix material has a large value of the product (density *times* specific heat), i.e. of $\rho_w c_w$ relative to the corresponding product (ρc) for the gas. The feature means in turn that the gas blows for a considerable time in a constant direction before the matrix temperature is significantly changed and the flow needs to be switched. Under these circumstances, gas mass throughput is many times the mass content of gas occupying the matrix. Put yet another way, the regenerator for furnace air pre-heating generally has a large *flush ratio*, N_{FL}, with numerical values in the thousands.

The equations which define the heat exchange processes and corresponding temperature variations are common to all cases. It follows that the method(s) of solution can be the same regardless of application. On the other hand, the extreme difference in operating conditions has been used to justify an analytical short cut when dealing with the air pre-heater (and, indeed, the Ritz type). The short cut involves proceeding from defining equations to solution *without* the term (a 'differential coefficient') which defines the flush phase. The rationalization goes that the flush accounts for a tiny fraction of the blow, and that the omission has an imperceptible effect on outlet temperature over the majority of the cycle. An exception to this practice is necessary when dealing in detail with the initial, flushing blow on start-up (see account below of the important treatment by Schumann).

It is the concept of flush ratio which puts orthodox regenerator theory into context. In the Stirling engine flush ratio, N_{FL} *always* has a small value – from less than unity to a maximum of about 3. Far from being incidental, therefore, the flush phase is all. So for the thermal analysis to assist with regenerator design, it must completely reflect the temperature response as the contents of the matrix are 'flushed' by fluid entering for the reversed blow.

In summary, solutions of regenerator thermal response in cyclic operation available to this stage intentionally ignore the flush phase. Solutions which cope with the flush phase do so only as far as the initial blow. Evidently, neither type of solution serves directly. There is nothing wrong with the underlying analysis – it just needs solving for the due emphasis on the flush phase. But there is a hurdle: Stirling engines, on the whole, attract engineers. The regenerator problem attracted analysts – a breed equally at home with transformations of the independent variable as with Bessel functions of the imaginary argument. Jakob's lengthy review of 1957 is skilful in combining a comprehensive treatment of the various mathematical approaches with readability if one wishes to skip the algebra. Thermal design of the Stirling engine regenerator calls for a different approach – not a compromise, but a solution method in keeping with the nature of the physical problem. The first step is an agreed picture of the thermal system under study.

5.5 A first physical model

Most analysts are happy with a somewhat stylized representation – at least at the stage of fixing ideas. Figure 5.2 shows the usual picture with the matrix contained within a duct of uniform cross-sectional area. A valve at the left-hand end can be switched rapidly between a port admitting a steady flow of gas at high temperature, T_E, and an exhaust. At the opposite end, a similar valve admits cold gas at temperature, T_C, in one position, and connects to exhaust in the other. This simplified picture of the furnace gas pre-heater does not, at first sight, correspond to conditions in the Stirling engine. The fact may help to explain why orthodox regenerator theory and the study of the Stirling engine have evolved separately.

By the end of the 1920s the problem of predicting the thermal behaviour of the innocent-looking system of Fig. 5.2 had attracted the interest of a number of eminent mathematicians in addition to Hausen. Mostly German, these included the giant of heat transfer, Nusselt (1927) and Schumann (1929). Schumann's name is not routinely associated with the Stirling engine, but his contribution was eventually to have a greater impact than that of any other analyst. He studied the first flush of a fluid at uniform temperature and flowrate into a duct initially at different, uniform temperature. This has since become a 'standard' analytical problem in its own right – the 'conjugate heat exchange problem', sometimes less pretentiously called the phenomenon of 'turning on the hot tap'. The latter description recalls the common experience at the hotel wash-basin or bath where hot and cold taps are not clearly marked. Whichever is turned on, the initial flow of water is cold. If the hot tap has been selected, the water temperature shortly increases – but gradually, reflecting an exchange of heat with the pipework. Eventually, both pipe and water achieve thermal equilibrium and maximum water temperature is achieved.

Figure 5.3 shows fluid and wall temperature relief converted from a Schumann solution covering the initial flush from entry to completion of penetration. The vertical axis is essentially proportional to temperature. The left–right axis is fractional distance through the matrix. The direction into the paper (labelled ϕ) is time as a fraction of the complete blow, t/t_{blow} (or, equally, crank angle). The upper figure is for the fluid, and records the rapid attenuation of the temperature front as heat is increasingly lost to the matrix. There is no change in exit temperature (right-hand end) until the initial flush is complete. The lower figure is for the matrix, showing the steady heating process free of any discontinuities.

It is important to a full understanding of the role of regenerator theory to appreciate that Fig. 5.3 is not as originally constructed by Schumann. The graphs are not even plotted in terms of the same variables. To give physical

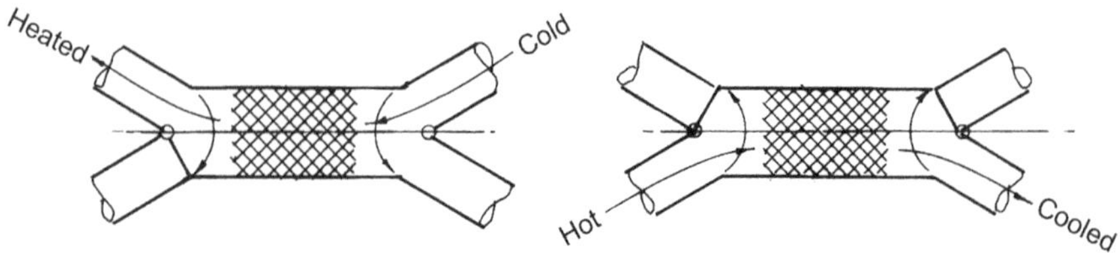

Fig. 5.2 Stylized representation of regenerator operation

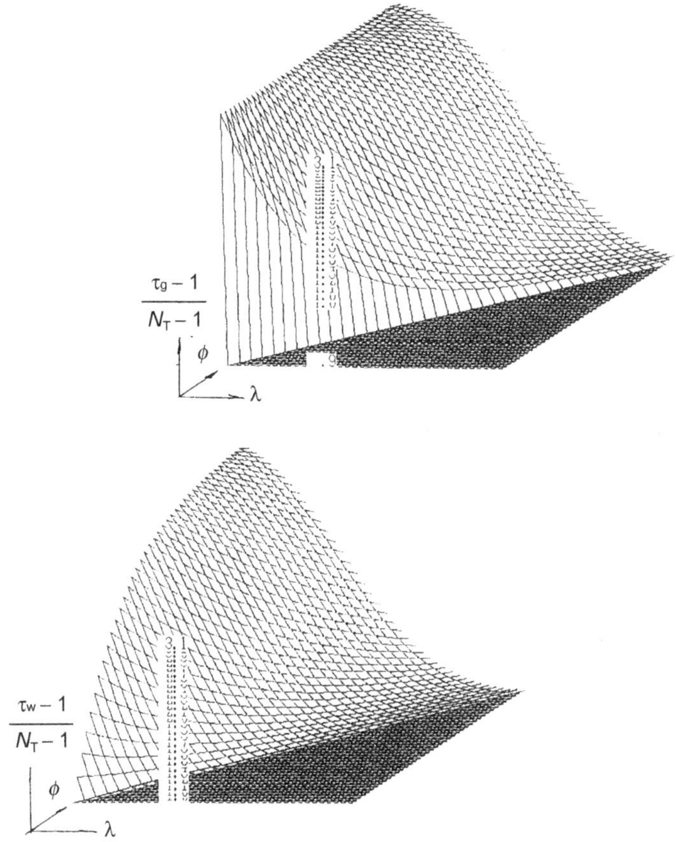

Fig. 5.3 Schumann's solution for the initial 'hot' blow transformed back to 'real' (*x*,*t*) coordinates and plotted in perspective. Upper panel, fluid temperature history, showing advancing step front. Lower panel, matrix temperature history

meaning they have been subjected to a 'reverse transformation' to the time–distance (x, t) coordinates of the underlying problem and replotted (Organ 2000a). In original format, the horizontal axis of the Schumann plots reflected a mathematical transformation from the x, t system in which the problem is posed to a new, cryptic counterpart in which the progress of the flush is *implicit* (not visible as such) rather than *explicit* as in the 'natural' presentation of Fig. 5.3. The present author, considering himself something of a dab-hand in regenerator theory at the time, memorably misinterpreted a plot (Kalinin and Dreitser 1994) of the cryptic sort, and felt the need to apologize. An embarrassing experience was not wasted, however: it served to reveal a cardinal reason (there are actually two) for the failure of orthodox regenerator theory to translate directly to the Stirling engine context: the inscrutable analytical procedures and notation. Who but a mathematician could roll together a group of parameters including heat transfer coefficient, *h*, density, ρ, specific heat, *c*, velocity *u*, combine with matrix length, L_r, and label the result *reduced length*, Λ; or make a marginally different combination with blow time, t_{blow} and call it *reduced period* Π? A return to the account of hardware development will assist in the evolving evaluation of the contribution of regenerator theory.

5.6 Back to the (Philips) Laboratory

The early phases of the Philips programme were dedicated to the construction and evaluation of engines to a wide variety of configurations. Several were constructed in an opposed-piston layout, reflecting the fact that, as far as the internal processes are concerned, this was the *'most favourable form that can be imagined'* (van Weenen 1947). Figure 5.4 is an example of the Type 7 variant dating from about 1939. The bore was 40 mm, stroke 30 mm and swept volume 75.4 cm^3. Pressurized to 98 atm with air, 300 W was claimed at 2000 r/min. Details of the regenerator are not available, but at the time Philips were using stacked pads of compacted matted nichrome wire of 30 μm. Problems reported (Hargreaves *op. cit.*) were with lubrication and with the brazed joints between heater and regenerator, but one suspects more fundamental limitations: the entire gas pressure is transmitted from gudgeon pin to gudgeon pin via the crank mechanism; the thermal conduction path between heater and cooler is short; there are two sets of piston seals under the full gas pressure differential. On the other hand, Philips had been doing considerable lateral thinking: an ingenious crank linkage gave the required phase angle between the volume variations. Whereas the coaxial, opposed-piston arrangement is nowadays the standard configuration by which the operation of the Stirling engine is analysed, there had at the time been no precedent for this thermodynamically attractive layout.

Philips' work on Stirling cycle machines during the German occupation is now the stuff of legend. It has been chronicled in too great detail elsewhere (e.g. by Hargreaves *op. cit.*), to bear repetition, but some highlights are relevant. The problem of raising mean working pressure without a corresponding increase in drive mechanism loading was dealt with by abandoning the opposed-piston layout in favour of a return to Stirling's original vertical, coaxial configuration. The resulting Type 10 engine has already been illustrated in Figs 4.1 and 4.2. It was still a true 'air' engine and pressurized by an integral pump to 5 atm, although it was apparently also the vehicle for the first serious investigation of the potential of hydrogen as working fluid.

5.6.1 An early approach to regenerator design

The Type 10 was the forerunner of the power unit of the beautifully engineered

Fig. 5.4 Philips Type 7 opposed-piston engine. Reproduced with permission of Philips Electronics BV

MP1002CA generator set shown in Fig. 5.5, and embodied a novel approach to expansion cylinder construction: a thin shell of alloy steel with a high creep-to-rupture strength at working temperature (600–650 °C) withstood the working pressure, while heat transfer area inside and out was enhanced by brazed-on fins of aluminium bronze. Regenerator construction was an innovation: the item was produced on coil-winding machinery from a continuous filament of fine, crimped wire, and filled the annular space between the (identical) aluminium–bronze heater and cooler elements. (If Philips cannot wind a coil, who can?) Volumetric porosity was later estimated (Organ 1997a) at 0.8, which tends to support Philips' objection to the inadequate 'fill-factor' of this method of manufacture. The critical point for present purposes, however, is that such a regenerator has a wide range of geometric parameters – wire diameter, pitch of winding, pitch, amplitude, and form of crimp, etc. – each combination promising different heat transfer and flow friction characteristics. To this day, heat transfer and flow properties for this form of construction are not documented in the open literature.

By 1944, however, Philips had developed an oscillating flow apparatus for exploring the thermal characteristics of regenerators. Regenerator thermal performance was (and still is) computed and measured in terms of a parameter called *Number of Transfer Units* or *NTU*. This sounds like a mathematical abstraction, but in fact captures the essence of steady, forced convective heat transfer in a single concept. To this extent, it is worth introducing it informally in advance of a more comprehensive definition later: under given conditions of steady flow, a heat exchanger (or regenerator) has a certain *capacity* for heat exchange by convection per unit temperature difference. The flowing fluid has a *capacity* for carrying away that heat per unit temperature rise. The *ratio* between these two capacities is called *Number of Transfer Units, NTU.* Another way of looking at *NTU* is as the grouping into which the parameters of the algebra automatically fall when one sets up the expression for the temperature distribution in steady, incompressible, convective heat exchange. Its numerical value is just a number, without dimensions, and is all that is needed to specify the distribution of fluid temperature between inlet and outlet.

The Stirling engine regenerator is *not* a steady-flow problem. However, by assuming thermal capacity ratio, N_{TCR}, to be infinite and focusing on the post-flush phase, Hausen had managed to obtain an approximate expression for temperature recovery ratio, η_T, in terms of *NTU* (his 'reduced length', Λ). Philips ingeniously linked the concept of η_T to *regenerator heat deficit*, ΔQ_r, defined as the difference between the heat carried away from the hot space by the gas leaving it, and that returned to the hot space by the gas on its return pass. If the total heat stored in the regenerator per cycle is Q_r then the relative deficit is $\Delta Q_r/Q_r$, which relates to $\eta_T = 1 - \Delta Q_r/Q_r$r as

$$\frac{\Delta Q_r}{Q_r} = 1 - \eta_T = \frac{2}{1 + NTU} \tag{5.1}$$

The seductive simplicity of the result is precisely in line with the designer's prayer. However, at the time of its use the regenerator problem had not been solved for the unsteady conditions in the Stirling engine, *and the effect of Hausen's drastic assumptions was unknown*. It will be of interest to return to this central point later. In the meantime, the performance breakthrough evident in the Type 10 and MP1002CA air engines appears to owe more to exemplary mechanical and metallurgical engineering than to any scientific progress with regenerator design.

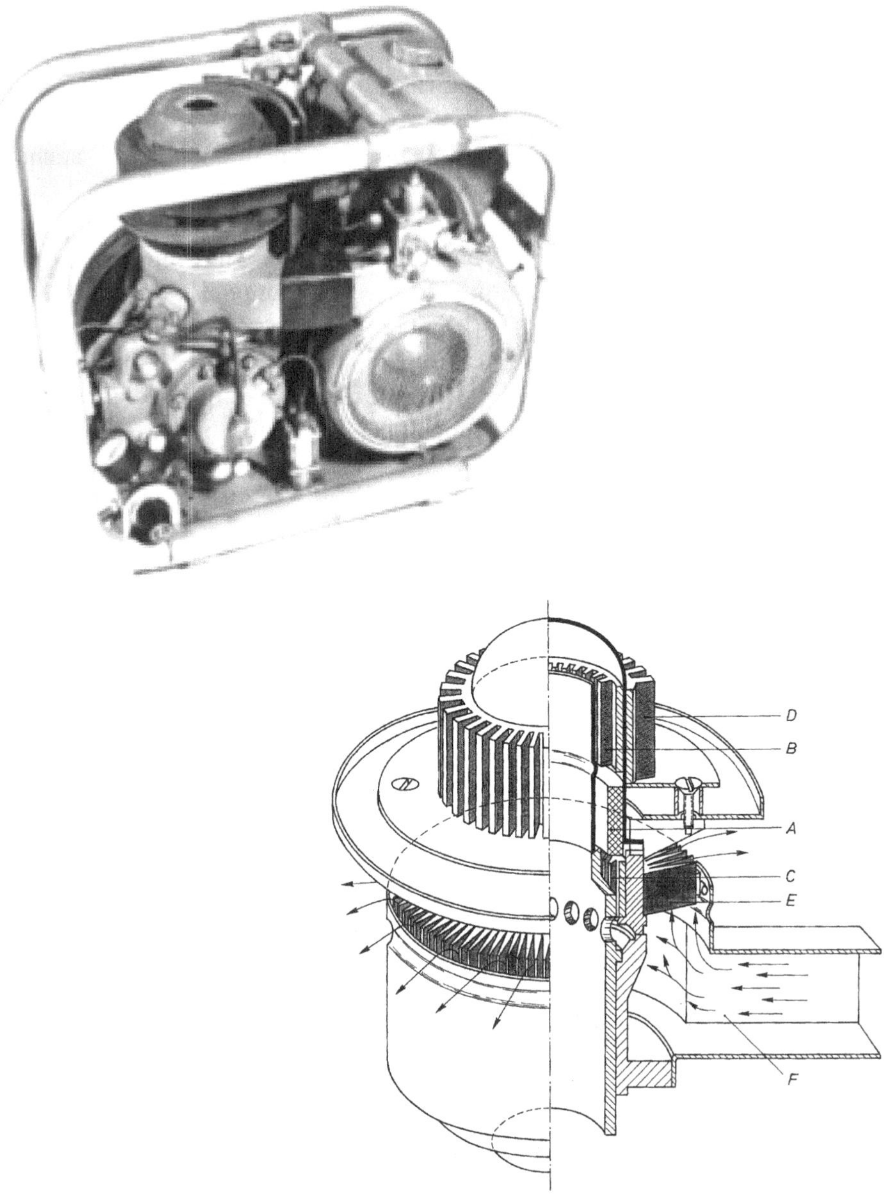

Fig. 5.5 Generator set rated at 200 W, designated MP1002CA and powered by a derivative of the Type 10 engine. This magnificently engineered product reached the batch production stage. The inset drawing shows the axial fins brazed to the pressure vessel of heat-resisting steel. Inset reproduced with permission of Philips Electronics BV

5.6.2 *Rebirth of the multi-cylinder concept*

Van Weenen's four-cylinder Type 19 engine of 1943 has been introduced in Fig. 4.3. It broke little new technological ground in the sense that the cylinder and regenerator units were essentially those of the Type 10 engine. The double-acting principle had been anticipated in a patent granted to Franchot in 1853. On the other hand, the Type 19 and its successors (Type 20 was a slightly enlarged version) took levels of specific power and reliability a step closer to those of the internal combustion engine. The importance to the evolving story of the air engine is that these models were essentially responsible for the spread of Stirling engine development work outside Eindhoven – and, indeed, outside Holland. Hargreaves (*op. cit.*) gives the details of these early collaborative efforts, including a telling account of SMF-Kroon venture. The significance of this latter engine for the present account is such that it is accorded a separate heading.

5.7 The SMF-Kroon engine

The engine was designed around the crank-case and crankshaft of a marine diesel engine. As shown in Fig. 5.6 this called for the gas circuit to straddle two adjacent, parallel cylinders. This was achieved by forming the expansion exchanger from tubes bent into loops. (Incidentally, it is difficult to agree that this was an *'original variant of the Rider configuration'* as claimed by Hargreaves, since the regenerator was now properly connected to the expansion exchanger. In the original Rider it connected to the *cold* end of the hot plunger.) The engine was a near-contemporary of the first Thermomotor engine. This had been a disappointment in terms of expansion exchanger performance, despite embodying the ingenious principle of the 'partition heater'. Whether the novel exchanger of the SMF-Kroon engine was a response to the Thermomotor problem is not known, but the multi-tube exchanger had arrived, and was here to stay. Prospects were set to change yet again – although perhaps not for the obvious reason.

To appreciate this, it is necessary to know that the power available from a heat engine depends on the mass of working fluid (e.g. air) it can process *per cycle* – and, of course, on the number of cycles per second (or r/min). A gas turbine with a 2 m diameter inlet ingesting air of density 1.2 kg/m^3 at a speed of 200 m/s processes some 700 kg – or three-quarters of a ton – of air per second! The figure explains the prodigious specific power, and why reciprocating engines will never compete. Doubling the mean pressure of a Stirling engine does indeed double the working fluid mass which can, potentially, be processed per cycle. To this extent it doubles potential specific power.

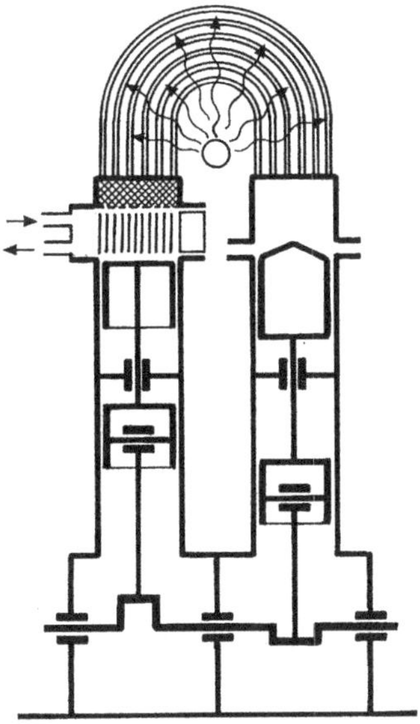

Fig. 5.6 Schematic representation of the SMF-Kroon engine. Reproduced with permission of Philips Electronics BV

On the other hand, the full benefits can be realized only if it is possible to double the original rates of heat supply and rejection. The tubular exchanger enables flow passage length and wetted area to be 'decoupled' from cylinder proportions, thus freeing up gas circuit thermodynamic design. However, the expedient has a price.

Adding tubular exchangers increases 'dead' (unswept) volume. All other things being equal, compression ratio is reduced, and on this count alone, the percentage power increase is less than the percentage increase in pressure. Put another way, working fluid mass has to be increased out of proportion. If a greater mass has now to be heated and cooled, yet further heat exchanger area is called for! The problem is exacerbated by the need to match the increased heat flux between tubes and working fluid with corresponding increases between heat source and the *outside* of the expansion exchanger tubes, and between the outside of the compression tubes and the heat sink. Relatively little problem is posed at the compression end, where the heat sink is usually a liquid providing a high coefficient of heat transfer. Things are different at the expansion end, where, in basic configurations, the combustion gases have one pass over a single – or at most, double – tube bank.

Returning to the original theme, the SMF-Kroon engine had been expected to produce about 50 h.p. at 500 r/min when charged with air to 50 atm. When measured output fell far short of this figure, a two-stage remedy was applied. Firstly the number of heater tubes was reduced by 23 *'and various other improvements made'*. Power increased to 31 h.p. at 370 r/min but problems of non-uniform heating remained, and there were *'temperature pulsations'*, particularly in the dead space on the compression side of the regenerator. The second stage of the remedy is credited to W. F. Schalkwijk, and consisted of cramming woven-wire kitchen scourers into all available dead space. Apparently power increased immediately to 45 h.p. at 500 r/min! Schalkwijk was Philips' regenerator analyst, whose contribution to regenerator theory is examined later.

Hargreaves admits elsewhere in his account of this phase: *'It was also clear that the agreement between theoretical calculations and measurements had to be improved!'* A dispassionate assessment of the prospects for the air engine might put things more bluntly: whatever the revival of the Stirling engine owed by this point to *'modern knowledge about heat transfer'*, etc., it is clear that the latter had yet to distil into a method of thermodynamic design which could cope effectively with the multi-tube gas circuit.

The same SMF-Kroon engine had the sad distinction of precipitating the next major change in direction when it exploded under load and caused a fatality. The investigation concluded that vapour or mist from the lubricating oil had entered the expansion space and reacted with the pressurized air. Nitrogen was immediately substituted, but it was soon decided that hydrogen and helium were preferable. After 1953 all Philips' designs were based on hydrogen or helium. The air engine became the hot gas engine and an era came to a provisional close.

The mid-1950s saw supersonic flight at 1132 mile/h (Orlebar 1990), the start of design of the Concorde supersonic airliner, and development of the Rover gas turbine-powered car. The concepts of thermodynamics, of convective, conductive, and radiative heat exchanger and of fluid flow which now form the core of the university-level engineering syllabus were all in place. The Method of Characteristics as an approach to problems in unsteady compressible flow was now a technique of some 60 years' antiquity (Abbott, 1966). Hausen's comprehensive treatise on the regenerator (*op. cit.*) had been in print for some years. A compendium of regenerator heat transfer and flow friction data (Kays and London) became available in 1955, and it is a tribute to its coverage and quality that it remains the definitive resource to the present day. This was no longer the age of caloric; in order to appreciate the implications of the new developments it is necessary to have a smattering of the new language. A brief digression will serve to bring the earlier, arcane notation of regenerator analysis into line with modern heat transfer practice, and to reveal the extent of Schumann's indirect, and largely unrecognized, contribution to Stirling engine technology.

5.8 Some basic concepts

There are a small number of concepts without which it is not possible to discuss – let alone to make sense of – convective heat transfer and flow loss. Algebra will be kept to a minimum consistent with the aim of building an accessible picture of the state of Stirling engine thermodynamic design.

5.8.1 The 'ideal' gas

The working fluid (air, helium, hydrogen) of the Stirling engine is invariably assumed to behave as an 'ideal' gas. This means that density, ρ [kg/m^3], pressure, p [Pa or N/m^2] and (absolute) temperature, T [Kelvin, K], are related through the *ideal gas law*.

$$p/\rho = RT \tag{5.2}$$

Where R is the *specific gas constant* for the gas under consideration. For air the value is 287 J/kg K. The reader might be interested in using the hand calculator to check that the density of air is about 1.2 kg/m^3 at ambient conditions (p = 10^5 Pa, T = 20 °C = 293 K).

5.8.2 Reynolds number

The Reynolds number, frequently denoted, N_{re}, derives from 1883 (White 1979) and is a dimensionless criterion of the *condition* of a fluid flow. Partly because it is dimensionless, it is independent of any specific fluid: a given, numerical value of N_{re} means a specific velocity profile – the same degree (or lack) of turbulence – whether the fluid be air, neon, helium or carbon dioxide.

N_{re} is defined in terms of variables density, ρ, mean mass velocity, $\underline{u} = m'/\rho A_{ff}$ [m/s], hydraulic radius, $r_h = A_{ff}/perim.$ [m] and coefficient of dynamic viscosity, μ [Pa s]:

$$N_{re} = 4\rho|\underline{u}|r_h/\mu = 4|m'|r_h/\mu A_{ff} \tag{5.3}$$

N_{re} has no units (no 'dimensions'). It takes only positive numerical values, the vertical bars bracketing a symbol denoting that the modulus (value without sign) is to be used.

5.8.3 Number of transfer units, NTU

Whereas the mathematician is happy to say 'let the coefficient of convective heat transfer be h [W/m^2 K]' the engineer needs to known *how many* Watts per square metre per unit of temperature difference. In other words, he or she needs to know the numerical value of the coefficient of convective heat transfer, h [W/m^2 K]. The information comes in terms of the geometry of the flow passage under consideration, and *the numerical value of* N_{re} *for the flow*. For economy and generality of presentation, h comes tied up with other parameters of the flow to make a dimensionless parameter, the Stanton number, N_{st}, defined as

$$N_{st} = h/\rho|\underline{u}|c_p \tag{5.4}$$

The physical significance of N_{st} is closely linked to that of NTU: whereas the latter was the parameter of a (steady-flow) exchanger in entirety, N_{st} relates to a local, representative element of that exchanger. It is the ratio of the capacity of the flow passage element for heat exchange by convection per unit temperature difference *and per unit area wetted* to the capacity of the flowing fluid for carrying away that heat per unit temperature rise *and per unit of free-flow area,* A_{ff}. It follows that N_{st} and NTU are related:

$$NTU = N_{st}\, L_r\, perim./A_{ff} = N_{st}\, L_r/r_h \tag{5.5}$$

NTU is identically the cryptic 'reduced length', Λ, of orthodox regenerator theory.

N_{st} takes only positive values. The relationship between N_{st} and N_{re} comes tabulated, or in the form of a graphical 'correlation' as in Fig. 5.7. Much early computer simulation work was in terms of specific value of h [W/m^2 K], back-calculated from the correlations in terms of numerical values for density, mean flow speed, $|\underline{u}|$, and specific heat, c_p.

It has since been appreciated that, as in the case of the steady-flow problem, there is economy of calculation and greater generality in formulating and solving for temperature in terms of N_{st} or NTU.

5.9 Schumann's solution to the initial blow

It is now possible to return to Schumann's contribution. To recognize the full impact it is necessary to know the following:

(1) The heat transfer coefficient (or NTU) of the regenerator cannot be measured in the same, straightforward way as that of the heat exchanger. This is because there is no way of continuously applying heat to the matrix so as to maintain it at uniform temperature, as the air (or other fluid) gains heat by flowing through it.

(2) Up until 1955 heat transfer correlations (i.e. tables or graphs of N_{st} versus N_{re}) for regenerator matrix materials were not available – at least, not in the open literature. Any regenerator temperature solutions available to this

> point therefore reflected *arbitrarily chosen values of heat transfer coefficient*. In other words, flow and heat exchange processes were represented as though completely independent – the heat take up of the fluid was not limited by flowrate, density, and specific heat.

This, of course, is the ultimate chicken-and-egg situation. However, thanks to the generalized nature of the Schumann solutions to the initial blow there is a most elegant way out. An experiment is set up in which the matrix under investigation is pre-heated to a uniform temperature by a steady flow of fluid. The incoming flow is then suddenly switched to a supply at the same mass rate but at a changed temperature. Matrix and fluid temperatures respond to this step change with a time-dependent change in fluid temperature at outlet – just as when the hot tap was turned on. In the Schumann solutions, outlet temperature for given volumetric porosity, flush ratio and thermal capacity ratio *is a function of NTU only*. The experimental output temperature record is compared with a range of Schumann solutions for a range of (arbitrary) *NTU* and the appropriate match is selected.

The resulting *NTU* value, of course, reflects the L_r and r_h of the particular gauze stack. The more general N_{st} value is retrieved algebraically as $N_{st} = NTUr_h/L_r$. In deference to Reynolds' analogy (a word about this in Chapter 6)

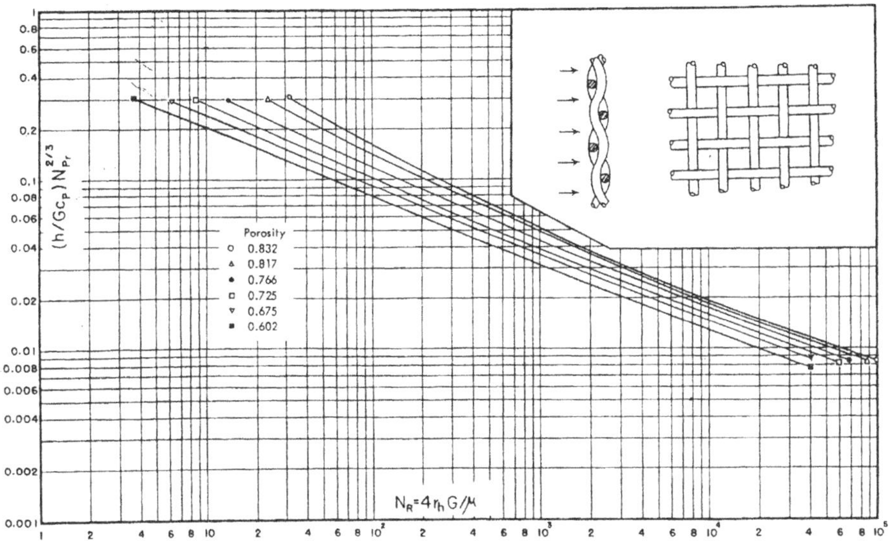

Fig. 5.7 Correlation of Stanton number, N_{st}, with Reynolds number, N_{re}, in the graphical representation pioneered by Kays and London. Reproduced with permission of McGraw-Hill Ltd and of Professors W. M. Kays and A. L. London

N_{st} is generally correlated with N_{re} as a product with Prandtl number, viz. as N_{st} $N_{pr}^{2/3}$ versus N_{re}. For present purposes the refinement is cosmetic: N_{pr} has a value close to unity for all gases of interest.

As far as is known, the first comprehensive account of the determination of N_{st} for wire matrix regenerator materials making use of the Schumann analysis is that reported by Coppage and London (1956), although the account draws on the dissertation by Coppage (1952) and other earlier sources. The N_{st} and corresponding friction data were organized into a first edition of Kays' and London's *Compact Heat Exchangers* which appeared in 1955. Figure 5.7 reflects the form of correlation already standard in the context of the isothermal, parallel exchanger (Chapter 6) showing both $N_{st}N_{pr}^{2/3}$ and friction factor, C_f, plotted against Reynolds number, N_{re}. Reassuringly, the curves are *not* parallel, the irregular passages continuously interrupt the flow: the celebrated Reynolds' analogy does not extend to this case. (Just as well in the circumstances that $N_{pr} \approx 1$ for candidate gases for the Stirling engine!!)

On the above account, 1955 may be taken to be the first public appearance of data basic to Stirling engine regenerator thermal design. By this date the revival at the Philips Laboratories had been underway for some 20 years, and the first hydrogen-charged rhombic drive engine, designated the 30-15 (Hargreaves *op. cit.*) had been designed and built – though not yet tested. However, this overlooks some relevant interim developments.

5.10 Interim summary

What has been the contribution to prospects to this point of *'modern knowledge about heat transfer, flow resistance?'* A look at the Stirling brothers' engine of 1845 reveals that it embodied a regenerator and was pressurized. Pressure variation is stated to have been between 10 and 15 atm, so the degree of pressurization was essentially that of Philips' first 'production' air engine – the legendary MP1002CA. Air entering and leaving the compression end of the displacer cylinder was forced over banks of coiled tubes containing cooling water, reflecting identically the intent of the modern counterpart. A cross-section through the displacer cylinder (Fig. 2.7) reveals channels cast into an internal liner to increase the area of hot metal surface wetted by the air flow – again anticipating precisely the function of the slots or fins of the redesigned engine. The contributions of the modern technological age *effectively embodied* in Stirling engine to this juncture were therefore:

(a) materials of high creep strength,
(b) brazed-on extended surfaces of high-conductivity material,
(c) high-speed operation allowed by the more compact crank mechanism,
(d) enclosure of the latter inside the pressurized body of the engine.

The use of finely divided wire for the regenerator can be thought of as an inspired choice, but only by coincidence could diameter, porosity, and layout have been those affording a favourable balance between heat transfer and associated flow loss. Indeed, there is nothing to confirm that a further, more elementary design balance had been struck, namely, that between the (inevitable) dead volume of the regenerator and its other functions. For given volumetric porosity, $¶_v$, dead volume, V_{dr}, is proportional to regenerator envelope volume. All other things remaining equal, an increase is *always* detrimental to compression ratio and thus to specific power. On this count the ideal is zero volume! The false conclusion admirably summarizes the challenge of Stirling engine thermal design: every single aspect interacts with all of the others.

Progress to the mid-1950s can be accounted for in terms not of science but of admirably sound, contemporary engineering practice. This compensated for earlier technological complacency and neglect by embodying interim developments in other heat engines and compressors. But the regenerator which equipped the first generation of twentieth century Stirling engines was not born of coherent scientific study. Indeed, some decades were to pass before it was generally felt that such study might be of benefit.

6

Reassessment

6.1 Status quo

More than 60 years have passed since the Stirling engine was rescued from oblivion at the Philips Laboratories. It has thus been under modern development for almost exactly the same time as the gas turbine. The latter had already entered military service by the end of the Second World War, and now powers every aircraft with the exception of trainers and light twins. The materials problems and the thermodynamic and fluid flow challenges have much in common with those posed by the Stirling engine. Yet with the exception of a small number of units generating auxiliary power in submarines (Bratt 2000), the Stirling engine, by contrast, remains at the laboratory prototype stage. Is it really so much more daunting a design and development challenge than the gas turbine? Or is the Stirling engine concept fundamentally flawed? Alternatively, is it just that something has not yet fallen into place?

The remainder of this book is dedicated to this last point of view – but there is evidently much explaining to be done. A catalogue of engine types – however, fascinating in its own right – will not serve this purpose. It is necessary instead to face a harsh reality: if an engineering system does not perform to specification, then (a) the design is wrong, or (b) the specification is wrong, or (c) both are wrong.

6.2 What is the Stirling engine design problem?

There are only two aspects of Stirling engine design which are unique, i.e., which cannot be dealt with by drawing on the technology of the gas turbine, the oil-free compressor, the internal combustion engine, etc. These are:

(1) Thermodynamic design. The sizing of heat exchanger and the specification of a regenerator to carry the required thermal load with minimum dead space and minimum resistance to gas flow.

(2) Context design. The proposed application (energy source) must complement, rather than strain, the design concept (choice of working fluid, etc.)

An instance of a poor match between context and concept is provided by the design study (Walker 1981) of steam rail locomotives converted to Stirling propulsion with power pistons direct-connected to the driving wheels through scotch yokes. The fact that the Stirling engine did not, and could not, be a success in this role reflects upon the *context/concept match* and *not* upon the Stirling engine. By the same token, embodiment of the sophisticated, vibration-free rhombic drive into a Stirling engine for domestic Combined Heat and Power (CHP) would amount to bad design if the capital cost premium meant that the payback period of the energy saving exceeded service life. The identical engine might pass as a piece of exemplary design if economics, together with technical performance, were favourable.

By the end of this book, it will have been comprehensively argued that these two, distinct aspects of design have joint responsibility for the fortunes of the Stirling engine. However, whereas responsibility is more or less equally shared, it takes a considerably greater amount of text to deal with thermal design than with the matter of concept versus context. A suitable treatment of the former should, however, take the latter in its stride, thus conveniently setting the format for the rest of this account.

Gas circuit design is not like the rest: drive mechanisms, bearings, and seals can all be 'seen'. Those aspects which are not visually obvious – loads, speeds, wear – can be measured by adapting techniques and measuring equipment widely employed elsewhere. The same is largely true of combustion chamber operation, where temperatures, flowrates and product concentrations can be monitored during operation. The internal gas processes stand in complete contrast: monitoring of cyclic pressure variation is feasible, but recording of fluctuating temperatures and flowrates inside the gas circuit is fraught with difficulty. Measurements from within the regenerator which convincingly distinguish between the temperature of the fluid and that of the matrix have yet to be presented.

The only way of probing these 'invisible' processes is by *inference*, and from a combination of approaches, including 'non-invasive' measurement, analysis, and computer simulation. If claims for the future of the air engine are in terms of its thermodynamic performance potential, then these claims are going to have to be evaluated in terms of concepts from the studies of thermodynamics, heat transfer, and fluid flow. This, in turn, calls for definitions. Some of these are unique to the Stirling engine, others are long-established in the fields of heat transfer, and fluid flow in general. This chapter puts the essential definitions and concepts as simply as possible.

6.3 Fundamentals of thermal design

An earlier classification (Chapter 1) distinguished between air engines operating on open and closed cycles. It was subsequently concluded that the future for the air engine lay with the closed cycle. The examples of air engines in this category have fallen, in turn, into three principal subcategories. Detailed discussion will be facilitated by having a label for each type. Predicting and comparing performance potential calls for some elementary algebraic notation.

Figure 6.1 shows the three categories schematically, though not in the sequence of historical appearance. The first is the opposed-piston type, frequently labelled 'alpha'. It is illustrated here in two-crank form with phase angle, α, between the cranks. It may equally be represented in a 'V' configuration with both big-ends connected to a common crank-pin, and with angle α between the cylinder axes. The 'alpha' configuration differs from the other two in the sense that α is both the *kinematic* phase angle (kinematic – '*relating to pure motion*', Shorter Oxford Dictionary, 1993) and also the *thermodynamic* phase angle.

Configuration 2 corresponds to Stirling's original engine of 1816. The designations 'beta configuration' and 'displacer configuration' are interchangeable. In contrast with the alpha type, where expansion space variations are controlled by the expansion piston (alone) and compression space variations by the compression piston (alone), in the beta type, compression space variations are controlled by displacer and work piston jointly. For this reason, the kinematic phase angle is not equal in this case to the thermodynamic angle, and receives a different symbol, β.

The third, 'gamma', category is also of 'displacer' type, but with work piston and displacer in separate but interconnected cylinders. Both displacer and compression piston are again involved in compression space volume variations, and kinematic (β) and thermodynamic (α) phase angles are again different. An essential difference between the power-producing potential of beta and gamma types arises from the fact that, in the latter, the strokes of the displacer and work piston do not overlap, so that minimum compression space volume always exceeds zero. The feature may be seen as an extra 'dead space', or unswept volume, which reduces compression ratio – and thus specific power potential – relative to the beta type of comparable displacements. A hybrid case between beta and gamma arises when in a beta machine the cylinder diameter for the displacer is not the same as that for the work piston. The implications for specific power have been investigated by Takase (1972). Machines commonly built to this 'beta/gamma' form are reversed Stirling cycle coolers.

A possible reason for the persistence of the gamma form emerges when one looks beyond the schematic representation. The price paid for the inherent thermodynamic advantage of the beta type is the relative intricacy of the crank mechanism. This must not merely deliver the required motion, but keep the side forces between displacer drive rod and piston to a practical minimum. The gamma configuration is readily achieved using two conventional cranks. A single crank serves if one cylinder axis is rotated relative to the other by an amount equal to the kinematic crank angle.

The term 'Rider configuration' (Hargreaves *op. cit.*) has been applied to the alpha category, but this raises a difficulty: reference to the section through the Rider engine of Fig. 3.12 will reveal that the regenerator connects to the expansion cylinder at the *cold* end, rather than at the hot as in Fig. 6.1. Before it enters the regenerator the working fluid must therefore pass through the annular gap around the plunger, although deriving some regenerative effect in the process. The feature compromises the usual functioning of the regenerator in the interconnecting duct, where it operates over a reduced temperature differential. The feature suggests that the Rider configuration ought perhaps to be accorded a special category. The matter is not pursued here, however, as the gas process cycle of the proprietary Rider engine appears to have little future.

6.4 Equivalence conditions

Apart from working space temperatures, T_E and T_C, the most important driving force for the gas processes in an air engine is the volume variation. There is obviously a wide range of possibilities, and the last thing one wants to cope with at the design stage is a different thermodynamic picture for every different crank mechanism and for each configuration – alpha, beta, gamma – and for every different size of engine as well.

The problem is eased by working in terms of volume *ratios*: *thermodynamic volume ratio*, κ, is defined, regardless of engine configuration, as the ratio of compression space volume excursion to expansion space volume excursion

$$\kappa = \frac{V_{c_{max}} - V_{c_{min}}}{V_{e_{max}} - V_{e_{min}}} = V_C / V_E \tag{6.1}$$

For the alpha configuration, the thermodynamic volume ratio and kinematic volume ratio are identical: the compression piston sweeps out volume V_C while the expansion piston sweeps volume V_E. Things are different in the beta and gamma arrangements, where the *kinematic* volume ratio, λ, is defined as

$$\lambda = \frac{\text{volume swept by work piston}}{\text{volume swept by displacer}} \tag{6.2}$$

The algebraic relationship between the thermodynamic parameters, κ and α, and the kinematic counterparts, λ and β was first stated by Finkelstein (1960a)

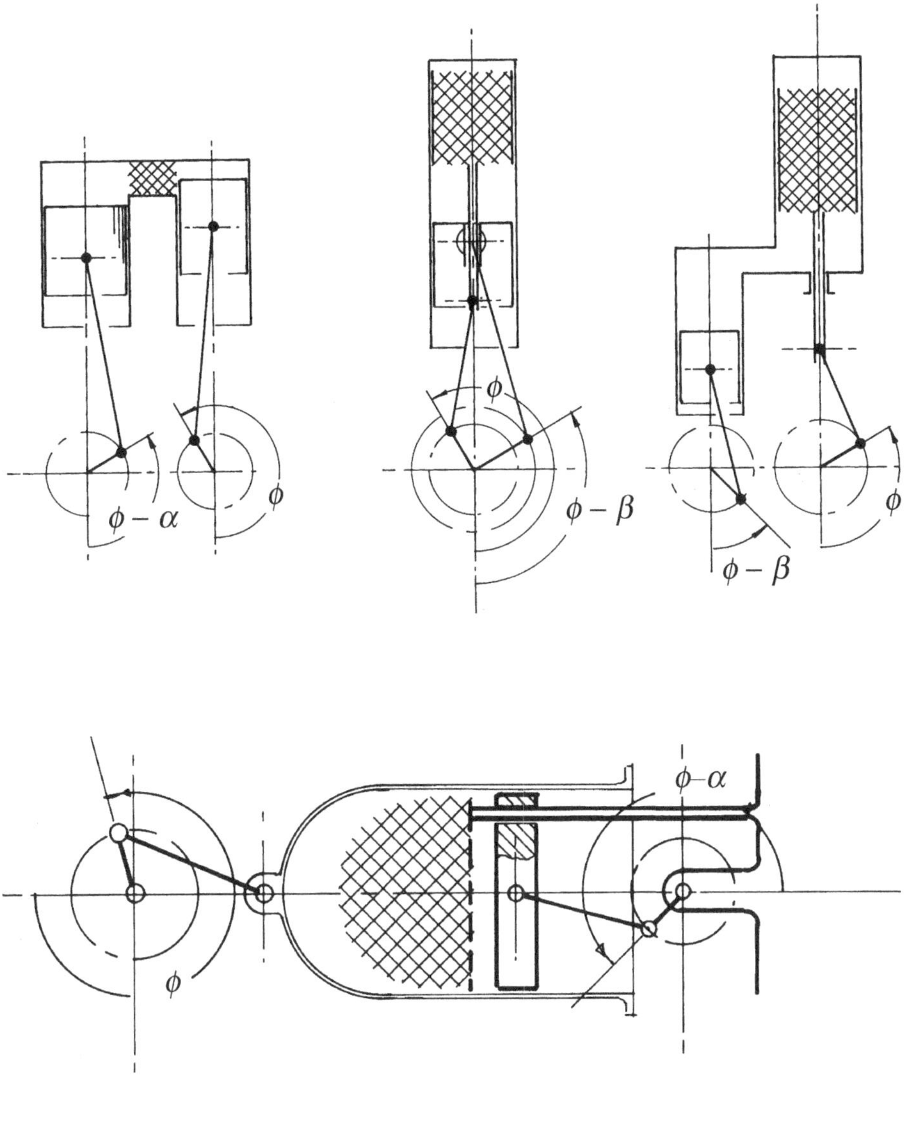

Fig. 6.1 Upper panel, classification of Stirling engines according to thermodynamic phase angle. Lower panel, hypothetical machine which confirms the equivalence relationships between opposed-piston and displacer machines by being simultaneously of both types

and permits the thermodynamic volume variations of a two-piston machine to be set to vary with crank angle in precisely the same way as those of a displacer-type machine (or, indeed, vice versa):

$$\tan\beta = \frac{\kappa \sin\alpha}{1 + \kappa\cos\alpha} \tag{6.3}$$

$$\lambda = \sqrt{\{1 + \kappa^2 + 2\kappa \cos\alpha\}} \tag{6.4}$$

$$\tan\alpha = \frac{\lambda \sin\beta}{\lambda\cos\beta - 1} \tag{6.5}$$

$$\kappa = \sqrt{\{1 + \lambda^2 - 2\lambda \cos\beta\}} \tag{6.6}$$

The work of inter-conversion is a matter of seconds on a hand calculator, but contains a potential hazard on account of the fact that the inverse trigonometric functions are multiple-valued. To find which value applies, the conversion may first of all be carried out approximately using the chart of Fig. 6.2. The result may, in any case, be considered sufficiently accurate for the purpose. If not, the relevant angle from the numerical calculation is that which lies closest to the value from the chart. (NB some calculators return angle in *radians*; to convert to degrees, divide by π (= 3.141592654 ...) and multiply by 180.)

The author has been involved in discussions on a number of occasions when doubt has been expressed about the equivalence conditions: is it *really* possible to design an opposed-piston machine in which the volume variations are *precisely* those of a displacer-type machine. Anyone who remains unconvinced may be interested in Fig. 6.1 (lower panel). The machine indicated is hardly a working proposition as an engine on account of the linkage connected to a point at high temperature, but in terms of the volume variations at issue it is both alpha and beta configuration in one. The κ/α notation is shown, but converts to the λ/β equivalent when the cylinder motion is 'stopped' by algebraically subtracting its motion from that of all the other components of the system.

The algebra to this point serves a further purpose: an alpha engine rated at 10 kW is not necessarily comparable with a beta engine rated at the same power until swept volumes, V_{SW}, are compared. V_{SW} for the beta type is equal to the volume swept by the work piston, in the alpha machine both pistons are involved:

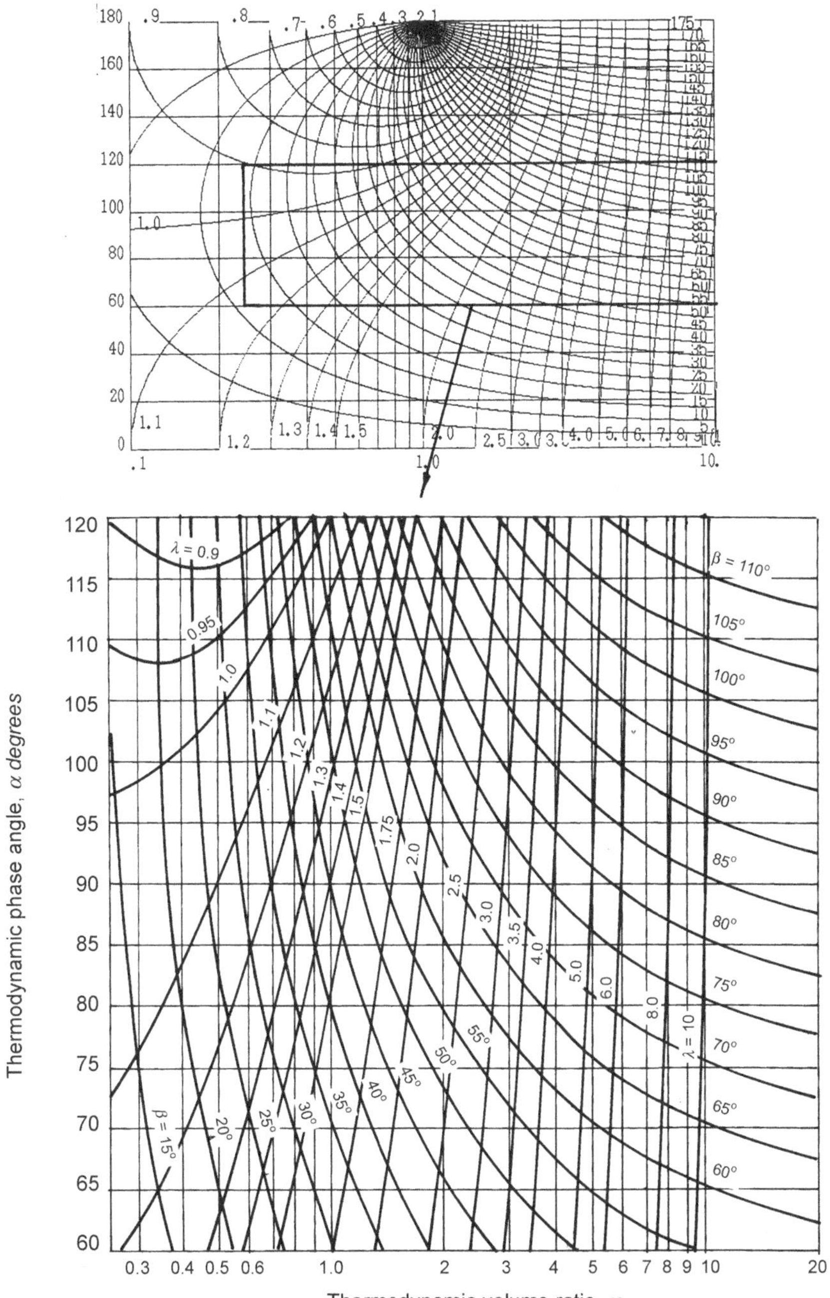

Fig. 6.2 Chart for inter-conversion between kinematic and thermodynamic parameters. The lower chart is an extract from the upper showing greater detail and is reproduced with permission of Cambridge University Press

$$V_{\text{sw(alpha)}} = V_E\sqrt{\{1 + \kappa^2 + 2\kappa \cos \alpha\}} \tag{6.7}$$

In equation 6.7, V_E is, as before, expansion space swept volume, which is identical with expansion piston swept volume. For the common case of $\alpha = 90°$ $(= \pi/2)$

$$V_{\text{sw(alpha)}} = V_E\sqrt{\{1 + \kappa^2\}} \tag{6.7a}$$

For $\alpha = \pi/2$ and $\kappa = 1.0$

$$V_{\text{sw(alpha)}} = V_E\sqrt{2} \tag{6.7b}$$

The United Stirling (USS) V-160 engine provides an example. It is of two-piston (alpha) configuration with a thermodynamic volume ratio of unity (coincident with kinematic volume ratio). Thermodynamic volume phase angle is the cylinder angle of the 'V' configuration, and is 90°, and expansion space volume, V_E, = 160 cm³. Swept volume, V_{SW}, is therefore $160\sqrt{2} = 226$ cm³. On this basis a more revealing designation would be USS V-226.

The intention in this book is to spread theory as thinly as possible, consistent with developing a useful picture of thermodynamic design. The account will accordingly be leavened with occasional digressions to exercise the various concepts as they occur. The vehicle chosen is Stirling's engine of 1818.

6.5 Reappraisal of the 1818 engine

It is reported (*Anon.* 1917a) that the engine produced 2 h.p. (1.5 kW) and pumped water from an Ayrshire quarry. If accurate, this would make it the most operationally successful prime mover of the genre to the present time. It is surprising, in view of the enduring fascination for the patents *per se* (Finkelstein 1959) that the engine itself has never been the subject of thermodynamic study – 'though the neglect is consistent with the persistence of an inappropriate cycle description through to the current and fourth edition of the UK's most celebrated thermodynamics text (Rogers and Mayhew 1992).

The balance will be redressed by reconstructing a thermodynamic specification of the engine using the tools as they are developed. The study quotes heavily from the author's paper (2000b) by permission of the Council of the Institution of Mechanical Engineers.

6.5.1 *Basic dimensional data*

The engine itself has long disappeared, and any attempt at reconstruction has to be based on the patent drawings in conjunction with the patent description.

It is of particular significance that the patent drawings from the Scottish patent (Fig. 6.3) are of near-photographic quality, with shadowing lending an element of perspective. If the finished engine was, indeed, in service in the following year (1818), then it may be speculated that design work was already complete or at least well-advanced. The draughtsman, possibly Stirling himself, may therefore have worked from design drawings and with the benefit of seeing castings or patterns. It is tempting to think of the drawings not as a concept sketch but a rendering of a thoroughly thought-through and engineered prime mover. The London patent, by contrast, has simple line drawings. The bell-crank connections are inverted relative to the Scottish version, reversing the direction of crankshaft rotation.

The scale of Fig. 6.3a (Fig. 8 of the specification) is stated to be *'one-half inches to a foot'* (1:24). The presence of the official wax seal prevents the making of one-to-one photocopies, but Miss Alison Lindsay of the Scottish Record Office kindly transferred strategic dimensions from the original to a bromide print. These give an overall height of 10 ft. Nineteenth-century English brickwork measures 3 in per course, and the height estimate is consistent with a count of the number of courses of bricks depicted in the masonry.

Coordinates of the drive mechanism have been picked off by scanning the print under a travelling microscope reading to ± 0.01 mm and normalized by crank-pin offset (semi-throw). Dimensions not available from the patent drawing are metal thickness, which, according to the specification, is *'to no scale'*. On the other hand, Stirling was explicit as to his preference for parts made by rolling sheet iron to be *'as thin as possible'*. He aimed for *'one-tenth of an inch'* (2.54 mm), and that this was the target thickness for the cylindrical sleeve between B and C.

The radial gap between displacer ('plunger') shell and the inside of the cylinder is given in the specification as a fraction (1/50) of diameter. This is the annular gap which was *'partially filled with wires wound round the latter and kept at a small distance from it and from one another by wires laid along it at right angles to the former ...'*. Wire diameter and material are not declared, but the possibilities are explored later in the course of this investigation. Later on we shall be in a position to dismiss the lack of information on the material of the regenerator wire as being irrelevant to thermodynamic performance in this particular case. In the meantime a preliminary idea of operating conditions is required.

6.5.2 Operating conditions

The source of the claim for 2 b.h.p. output has been traced back to an article in

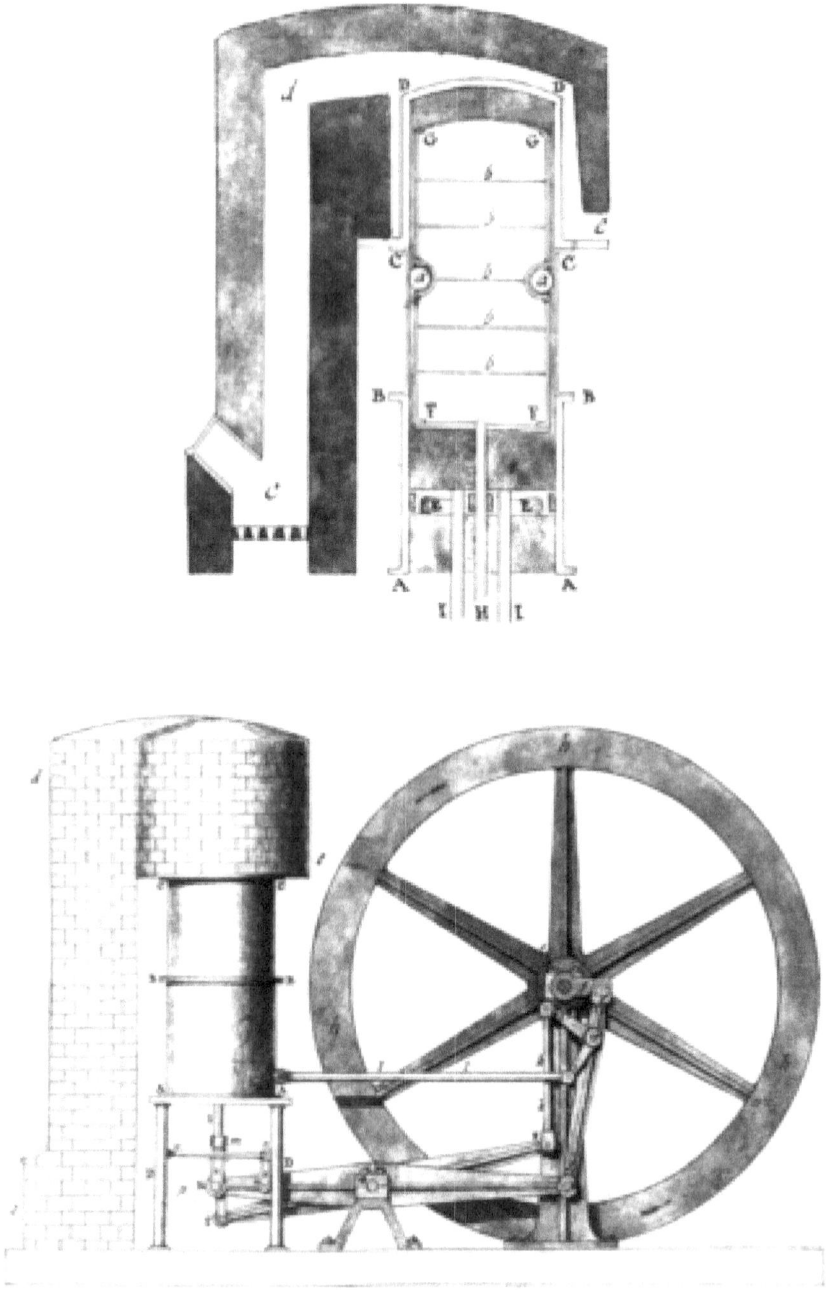

Fig. 6.3 Stirling's Figs 7 and 8 from the Scottish patent. Reproduced with the permission of the Keeper of the Records of Scotland (Reference C20/17/1)

The Engineer of 1917 celebrating the rediscovery of the patent specification – but no further. Neither is the value of r/min known at which the power was achieved. An account presumably exists, because James Stirling (1853) was able to report that *'it performed the work very well until, because of carelessness by the engine's operator, the bottom of the air vessel became overheated and was crushed by the pressure of the heated air'*. If the 1818 engine was constructed according to the patent drawing, then it was the *top* of the air vessel which was vulnerable to overheating. Moreover, the firing provision would make such overheating difficult. One is left wondering whether reference to the *bottom* of the air vessel is not perhaps a confusion with one of the later engines where heating was direct and from below.

This study could continue with angular speed left as a variable, but insight results from an attempt to bracket the actual value. A limit on the r/min of early prime movers was set by flywheel integrity. The flywheel in this case is an impressive 10 ft (3 m) in diameter. The slender, straight spokes shown in both patent drawings strongly suggest that this is not a solid iron casting – possibly Finkelstein's justification in 1961 for showing an assembly of bolted or riveted sectors, despite the fact that the feature does not show up explicitly on the drawings of either of the patent specifications. An elementary calculation shows flywheel rim hoop stress to be proportional to (diameter × r/min)^2. The drawing of Stirling's later engine whose rated speed is known to have been 28 r/min shows a flywheel of similar design but of 19 ft diameter. Equating the product of diameter with r/min for the two cases indicates 52 r/min for the 1816 engine if flywheel integrity was the limiting factor in both cases. An allowance for the inferior technology of the earlier flywheel suggests a value reduced in the direction of, say, 30 r/min. Later on, two independent approaches will be used to re-examine this result. Provisionally the study settles on a nominal 30 r/min.

6.5.3 Kinematics and volume variations

A first stage in the reconstruction of the thermodynamic personality of the engine has been to build a geometric and kinematic model (Organ 2000b) using ProEngineer™. This affords the opportunity of seeing this important prototype from any desired viewing angle, as in Fig. 6.4. More importantly, the modelling package permits the drive mechanism to be cycled through a revolution of the crankshaft, permitting verification of the kinematics defined by the data in Table 6.1.

The motions of piston and displacer create the volume variations determining the flow processes, and ultimately the indicated cycle work – which brings us back to phase angles and volume ratios.

The algebra of the drive kinematics is comprehensively developed in an appendix to the full account (Organ 2000b). The algebra has been coded

Fig. 6.4 Pro-Engineer rendering of Stirling's 1818 engine based on the data of the Scottish patent. After Lo (1998) with permission.

separately from the ProEngineer version for linking with existing gas circuit simulations. The algebraic notation and the first steps of a complete revolution of the crankshaft are illustrated in Fig. 6.5.

Figure 6.6 plots the motion of piston and displacer over a complete cycle. Being generated by an intricate mechanism, the motions are not simple harmonic, but phase angle and volume are the only convenient criteria, so we accept an approximation: *thermodynamic* phase angle, α, from the motion diagram may be taken as the angular difference between respective maxima of expansion space volumes. It may also be taken as the difference between the respective *minima*, in which case the numerical result is likely to be slightly different; some would recommend settling on the average. *Piston* phase angle, β, in contrast, is the angular difference between respective top-dead-centres (or, correspondingly, bottom-dead-centres.)

For this parallel-bore machine, thermodynamic volume ratio, κ, may be

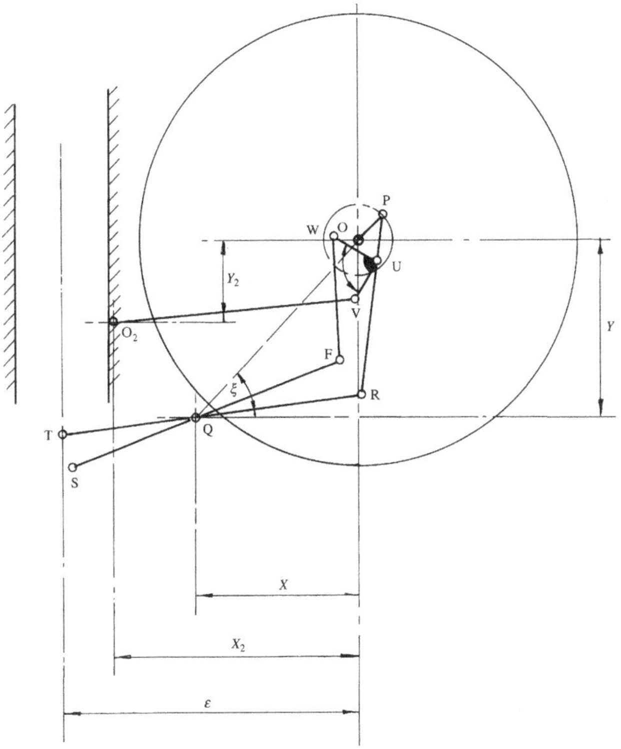

(a) Skeleton of drive mechanism with notation for Table 6.1

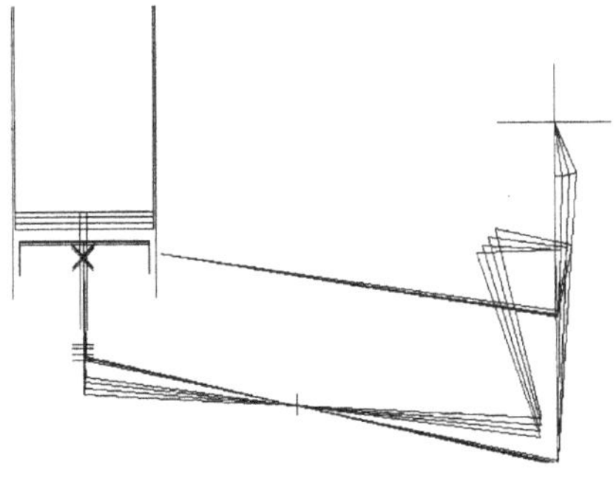

(b) First three increments in computer-animated linkage motion

Fig. 6.5 Reconstruction of the drive mechanism (NB: Scottish patent)

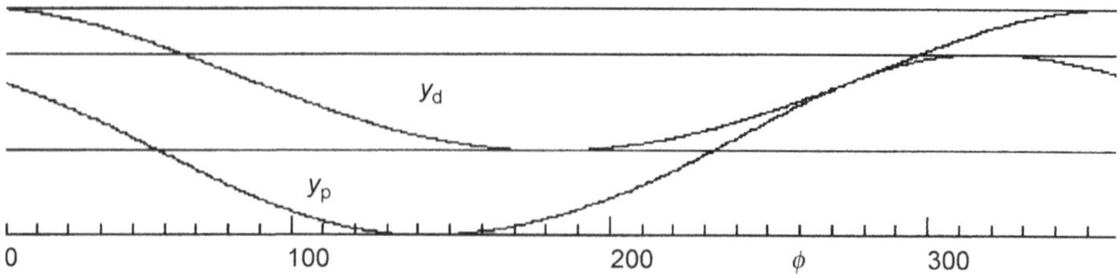

Fig. 6.6 Trace of piston and displacer motions of the 1818 engine (Scottish patent)

calculated as the ratio of maximum separation of the opposing piston faces in the compression space to the maximum vertical excursion of the displacer. Values scaled from the motion diagram add the values of piston and displacer stroke, S_p and S_d, to the specification of Table 6.1. Kinematic volume ratio, λ, is the ratio of respective vertical travels of piston and displacer.

The reader is invited to pick off numerical values of κ, α, λ, and β as suggested above, and then to use Fig. 6.2 and/or equations 6.3–6.6 to convert between one pair and the other.

The values should be close to those of the thermodynamic specification now taking shape in Table 6.1

6.5.4 Temperature ratio

A further, basic parameter of performance, both specific power and thermal efficiency, is *temperature ratio*. This is defined subsequently by the symbol N_T as

$$N_T = T_E/T_C \tag{6.8}$$

In equation (6.8), T_E is nominal expansion temperature in K, i.e. reckoned from absolute zero of -273 °C. T_C is nominal compression space temperature with the same datum. The operating temperature difference of the 1818 engine is declared in the specification as 480° (presumably Fahrenheit) or 480 × 5/9 = 267 °C *difference*. On the assumption of 30 °C at the compression end, this converts to a T_E of 297 °C, contributing a temperature ratio, $N_T = T_E/T_C = (297 + 273)/(30 + 273) = 1.88$ to the evolving specification (Table 6.1). Pressure is stated to fluctuate from a minimum below atmospheric pressure to a maximum above with a mean determined at the maximum power condition by leakage, and by leakage plus flow through a waste valve under reduced power. Only full power operation (waste valve closed) will be pursued here, with reference pressure (minimum, maximum, or mean) still to be estimated.

6.6 Some essential basics

6.6.1 *Significance of temperature ratio*

On the assumption that air behaves as an ideal gas at the pressures and temperatures involved, efficiency and power-producing potential indicated by basic cycle analysis are unchanged by changes in T_E and T_C which leave the value of N_T unchanged. The combination of 479 °C for expansion temperature with 127 °C for compression temperature would lead to the same N_T value (try the arithmetic) as that of the 1818 engine, and, on this basis, to the same performance. In reality, thermal conduction losses (thermal 'shorting') would be increased, and there would be other changes.

6.6.2 *Dead space ratio*

Discuss gas circuit design and sooner or later the matter of dead volume, V_d, will arise. Obviously, V_d and swept volume, V_{sw}, tend to vary in proportion, so the only meaningful number for design and comparative purposes is a ratio – *dead space ratio*, δ. There is a value relating to each of the two variable-volume spaces, and one each for expansion and compression exchangers (if the engine has exchangers – the 1818 engine did not), and one for the regenerator. Dead space ratio, δ_r, for the regenerator, for example, is defined in terms of pore volume, V_{dr}, as:

$$\delta_r = V_{dr}/V_{sw} \tag{6.9}$$

Expressions and values for other elements of dead space, e.g. that for the expansion space, δ_e, follow by analogy.

6.6.3 *'Extra' dead space*

All other things being equal, increased dead space means lower compression ratio and lower specific power. On this account it is important to note that, of the two displacer configurations, the parallel-bore beta and the gamma, the latter inevitably involves greater compression-end unswept volume, δ_c, than the former. The extra is over and above that of the interconnecting duct. It is due to the fact that, for viable kinematic phase angle, β, the two parts of the split compression volume are never simultaneously at a minimum. This additional dead space, δ^+, may be expressed (Finkelstein 1960a) by

$$\delta^+ = \tfrac{1}{2}(\lambda + 1 - \kappa) \tag{6.10}$$

The numerical value of equation (6.10) is evidently 0.5 for values of λ and κ each close to unity, meaning that the additional dead space is one-half of swept volume. As no designer of sound mind would intentionally introduce such a performance penalty, a preference for the lambda configuration over the beta calls for justification in terms of some considerable design benefit elsewhere.

A way of expressing the goal of gas circuit design is to see it as an attempt to *minimize* the various δ values while *maximizing* heat transfer area and minimizing flow loss. This difficult balancing act is beyond any explicit thermodynamic theory. The alternative – an experimental search for the optimum – threatens an infinite number of trial designs. If anyone tells you that he or she has 'optimized' a gas circuit design, nod politely and reserve judgement [but take a look at Chapters 16 and 17 of the author's text (Organ 1997a)].

6.7 Summary of fundamentals to date

The Stirling engine will not be ubiquitous like the internal combustion engine until it is understood as well as the internal combustion engine. In design 'understanding' means the intuitive 'feel' which short-circuits the need for computation.

For the purposes of the evolving picture, temperature ratio, N_T, thermodynamic volume ratio, κ, and thermodynamic phase angle, α, together with the set of dead-volume ratios, δ, are *fundamental design parameters*. The full significance will fall into place later. For the moment it is sufficient to imagine two Stirling engines, differing in terms of size, or speed, or charge pressure, or working fluid – or in terms of all of these. Both can be arranged to have essentially the same history of temperature–time–location and of pressure–time–location over a cycle *provided that* N_T, κ, α, *and the set of* δ *are the same for both.* Conversely, the gas processes in two engines for which respective numerical value(s) of one or more of these parameters differ are themselves necessarily different.

There are, of course, further design parameters; these will be deduced later. The N_T, κ, α, and the δ values however, provide the basic and necessary condition without which strict similarity of the gas processes is not possible.

Table 6.1 Basic data from patent specification (Scottish version), symbols refer to Fig. 6.5a

Working fluid	*atmospheric air*
Mean charge pressure	*10^5 Pa*
Swept volume	*0.103 m^3 (103 litre)*
r/min (to be verified)	*27*
Kinematics (Fig. 6.5a)	
Bore/stroke (D/S_p)	1.72
D/r	2.80
L_r/D	2.0
d_{rod}/D	0.071
X/r	4.93
Y/r	5.20
PR/r	5.33
RQ/r	5.08
α_{bc}	90°
PU/r	1.36
X_2/r	7.53
Y_2/r	2.4
WU/r	1.48
O_2V/r	7.67
UV/r	1.23
FQ/r	4.67
FW/r	3.60
SQ/r	4.06
TQ/r	4.13
e/r	9.0
α (equiv. volume phase angle)	134°
κ (equiv. swept volume ratio, V_C/V_E)	0.629
β (piston phase angle)	38°
Λ (piston volume ratio)	0.7826
L_{ref} (= $V_{sw}^{1/3}$)	0.469 m
A_{ref} (= $V_{sw}^{2/3}$)	0.22 m^2
A_{ffr}	0.018 m^2
Temperature ratio, $N_T = T_E/T_C$	1.78
¶$_v$ (regenerator volumetric porosity)	0.8 (assumed)
δ_r (at ¶$_v$ = 0.8)	0.221
ν_D (at ¶$_v$ = 0.8)	0.164
N_{MA} (dimensionless speed)	0.00451
N_{SG} (Stirling number)	2.08×10^9
N_{TCR} (thermal capacity ratio)	10 000

7

Post-revival

7.1 Synopsis

The fortunes of the Stirling engine subsequent to the original revival are largely in terms of the legendary rhombic drive engine, and of substantial programs of collaborative development between Philips, United Stirling Sweden, General Motors, MAN/MWM, Ford, and MTI (Mechanical Technology Inc.). There have also been independent research and development projects. The latter have generally been to a smaller scale, but the contributions have been out of proportion to respective budgets and manpower levels.

The digital computer emerged in the 1960s as the key to understanding the cycle gas processes. Cycle theory and simulation evolved more or less in parallel with engine development, but the open literature is confined to reports of the findings of the independent researchers who, on the whole, were not in the best position to compare the predictions of theory with experimental measurement. Possibly for this reason there was little evidence at any given moment of a beneficial influence on gas circuit design.

For technology to flourish the insight yielded up by individual advances has to be distilled to its essence and made accessible for others to build on. If each researcher has to replicate the work of every other, progress is minimal. The basic requirement for a technology to advance, therefore, is *communicable insight*. The failure of analytical and numerical studies to yield this contribution to thermodynamic design would be its most striking feature but for a still more remarkable aspect. By 1960, the literature on the core component, the regenerator, had already advanced in scope, depth and sophistication beyond the total output on the Stirling cycle to the time of this present writing. Yet until very recently (mid-1990s) no significant attempt has been made to bridge the gap between the two sciences.

As gratuitous criticisms these observations have little value. For an assessment of the prospects for the Stirling engine to have substance, it needs to reflect those avenues which have been exhausted and those which have not. The sections which follow will illustrate that the science of gas circuit design remains, nearly two centuries after the original invention, in its infancy.

The current predicament suggests a fundamental distinction between the regenerative thermal machine and almost all other cyclic thermodynamic systems: it is one device where *intuition alone* has failed as a method of design.

The numbers, lengths, and hydraulic radii of the exchangers must be determined in relation to working fluid and operating conditions (pressure, r/min, power level). This is not to say that the work of systematic design need be onerous. Indeed, it need not even call for use of the computer. For the process to become refined into a routine operation, however, the focus must increasingly be on the generation of *communicable insight*.

An illustrated account of the full, post-revival era of the Stirling engine is beyond the scope of this text. Attention from this point will accordingly be restricted to the air engine and to its fortunes and prospects. Nitrogen- and air-charged machines will be treated synonymously. Helium- and hydrogen-charged machines will feature, and prominently, but only to the extent that the design points under discussion have a bearing on the air engine.

7.2 The rhombic drive engines

The rhombic drive (Meijer 1960) was apparently conceived in the context of a small presentation engine to mark the retirement of H. Rinia. The feature had previously featured in a horizontally opposed internal combustion engine (Lanchester 1898), but Meijer's adaptation to the coaxial Stirling engine was original and inspired. It allowed the Philips programme to take off again with renewed vigour.

The mechanism is seen in Fig. 7.1. It consists essentially of two identical, counter-rotating crankshafts synchronized by gears. The upper set of connecting rods attaches to the piston, the lower to the displacer drive rod. The kinematic symmetry about the vertical axis means that horizontal forces, both static and dynamic, are in balance. Piston and displacer have purely vertical motion, so all components of side thrust are eliminated. If piston and displacer masses are equal, their combined centre of gravity moves along the vertical axis with simple harmonic motion, allowing perfect counterbalance by masses rotating with the crankshafts. Needless to say, analysis of the kinematics of the mechanism reveals that piston and displacer motions cause compression to take place largely in the compression space and expansion to take place largely in the expansion space as required.

The potential of the single-cylinder rhombic drive engine appears to have been the attraction when General Motors began collaboration with Philips in 1958. The Allison division of the American corporation contracted with the US Air Force to supply a self-contained generator set to produce 3 kW(e) from solar radiation when installed in a space satellite. The application highlights yet another unique feature of the rhombic drive: if the two counter-rotating crankshafts and associated flywheels and twin electrical generators are identical and mounted symmetrically, there are no overall gyroscopic effects.

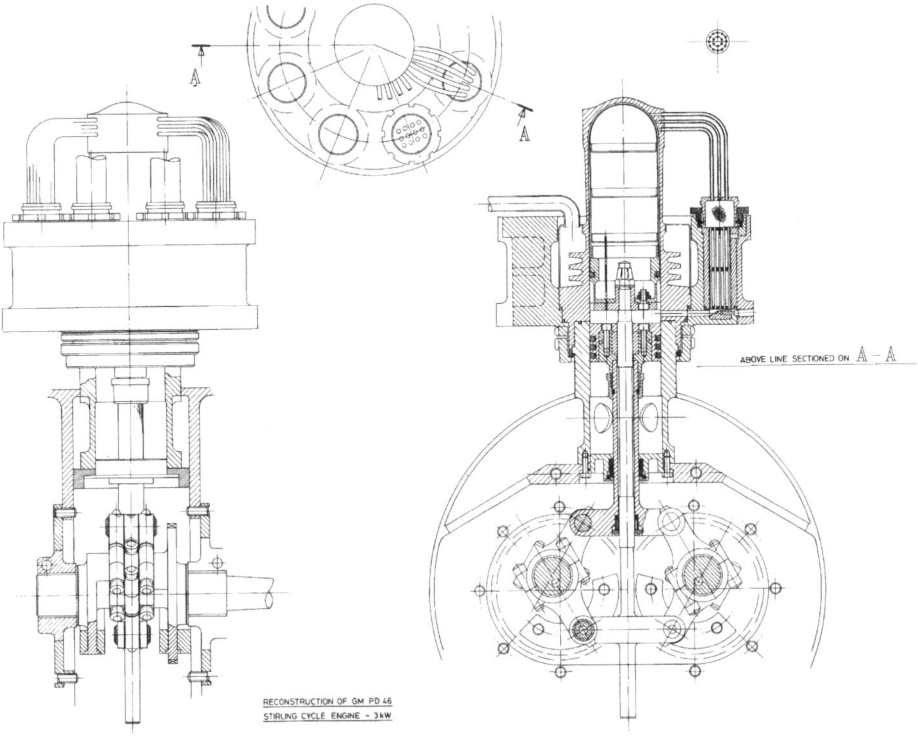

Fig. 7.1 General Motors' PD-46 engine for satellite power generation exploited the unique characteristics of the rhombic drive mechanism. After Organ (1997a) where the drawing is reproduced to a larger scale and accompanied by a detailed specification

The prototype space engine was a development of the Philips 10-36 model, and was designated PD-46 (PD for Philips Design) modified for heat input from molten sodium–potassium (NaK) eutectic which was to transfer the heat from the solar receiver. The principal features of the drive mechanism and the gas circuit are shown in Fig. 7.1. Tests using helium and nitrogen as working fluids led to the conclusion (Monson 1961) that performance with helium would benefit from an increase in Reynolds numbers in the expansion exchanger. When the engine was redesigned (as the PD-67) for reduced weight, the number of heater tubes was reduced from 96 to 76 and the inside diameter was reduced from 0.072 to 0.06 in (1.829 to 1.524 mm). Tube length was increased at the same time from 4.1 to 6 in (104 to 120 mm). The wider implications of the changes are discussed in detail elsewhere (Organ 1997a), but what is relevant here is that the 'rationale' for the modifications comes across in the reports not as part of a considered methodology of thermodynamic design, nor even as a product of computer simulation, but as completely *ad hoc*. If this impression is correct then the early 1960s see Stirling engine development still as a money- and time-consuming process of cut-and-try.

This is not to deny that progress was achieved: work for the satellite generator was overtaken by the development of electronics components with drastically reduced power demand, but attention transferred to the development of a power unit for military use having low detectability. The result was the GPU-3 (Ground Power Unit-3); Figure 7.2 shows the essential features. A GPU-3 eventually underwent trials in a hybrid (Stirling/electric) car. The General Motors' programme was extensive, embracing the design of a four-cylinder engine. Eventually, however, other priorities prevailed, and Stirling engine development ceased.

7.3 Sealing

The achievement of linear motion cried out to be complemented by an 'absolute' gas seal. The inspired response at Philips was to adapt the 'roll-sock' or rolling-diaphragm seal illustrated in Fig. 7.3. In principle, this is little more than a length of polyurethane tubing folded back on itself, but in practice calls for a supply of pressurized oil on the buffer space side to counterbalance the gas pressure. Reliability then becomes dependent upon that of the oil supply and pressure control. Moreover, fatigue life of the flexural element was apparently sensitive to temperature; the promising results of laboratory testing could not be achieved in engines in service, and an elegant concept was abandoned.

The matter of sealing is so 'obvious' that it tends to pass without comment, but it impinges on thermodynamic performance. Compression ring leakage has two consequences: parasitic pumping work (it takes energy to force a gas to flow across a pressure differential) and reduced compression ratio; both are to be avoided. Leakage of hydrogen and helium per unit pressure gradient is considerably greater than that of air (or N_2). Moreover, experience with miniature, reversed-cycle cryogenic coolers suggests that wear rate is high in an atmosphere of dry helium. Conversely, sealing problems of all types are less critical when air is used: had the rhombic drive engines been air charged, the rolling-diaphragm seal might have been unnecessary.

Most seals are of the 'rubbing' variety. The compression rings of early air engines such as the MP1002CA were self-energized metal rings running against an iron cylinder liner and were oil lubricated. The arrangement called for the lowermost ring to be a scraper ring, and both types of ring give the appearance of having been off-the-shelf automotive components. The small

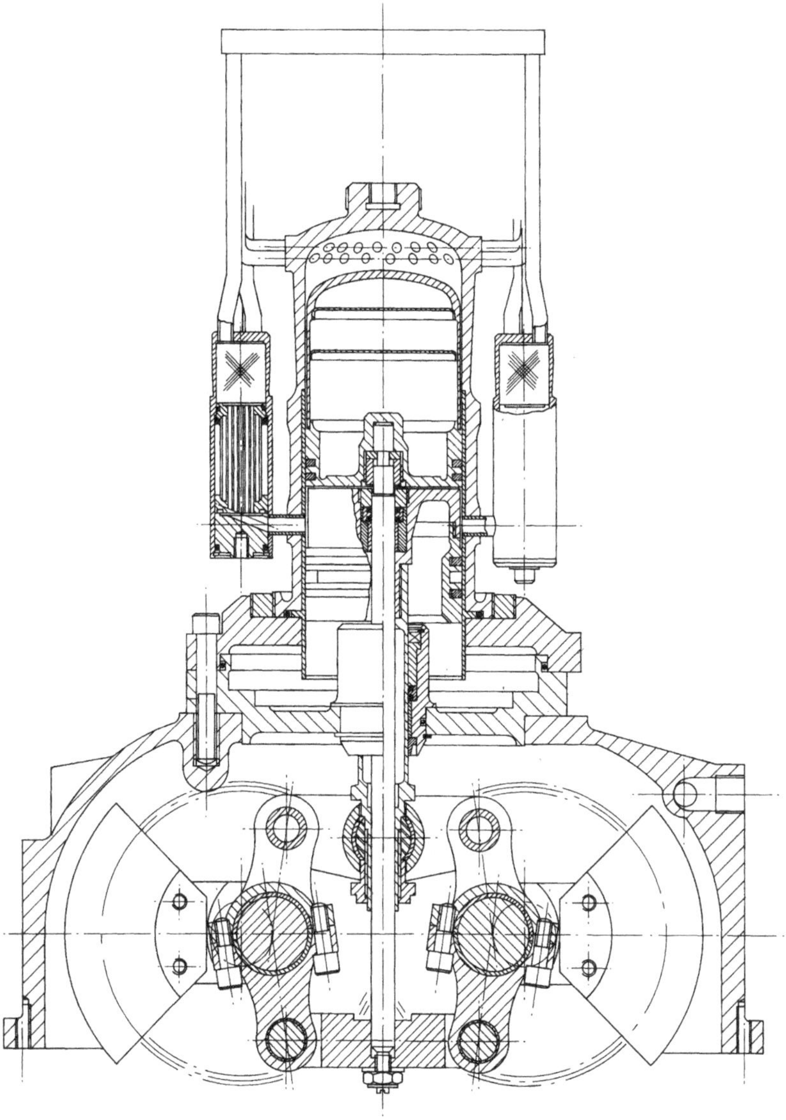

Fig. 7.2 General Motors' GPU-3 engine in cross-section. After Organ (1997a) where the drawing is reproduced to a larger scale and accompanied by a detailed specification

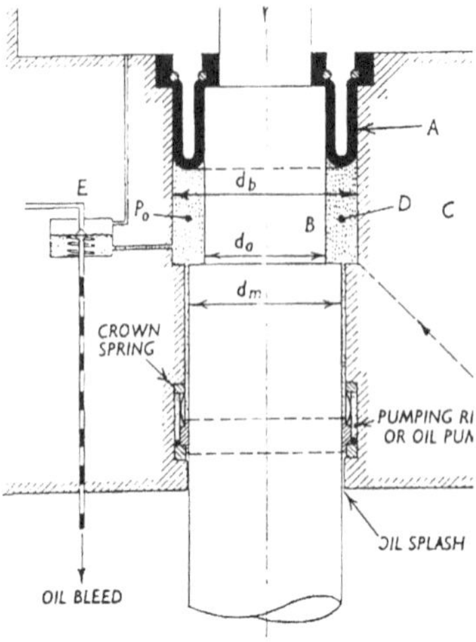

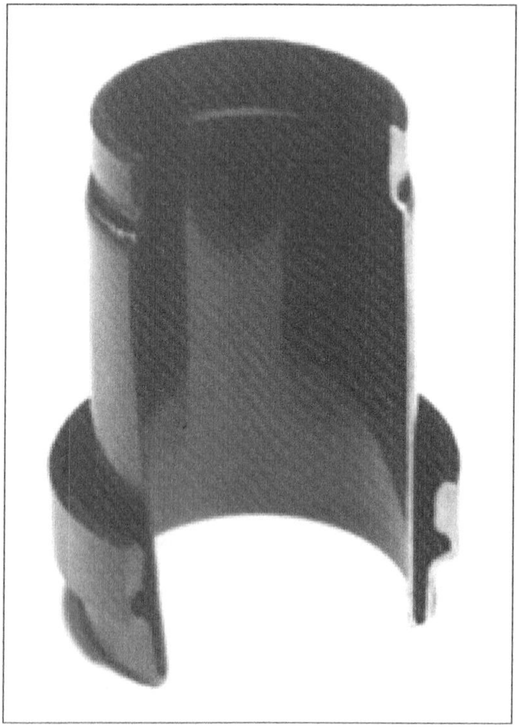

Fig. 7.3 The rolling-diaphragm hermetic seal. Courtesy of Philips Electronics BV

pressure difference across the coaxial displacer allowed use of a plain (clearance) seal assisted by marginal oil lubrication.

More recent designs have used dry-running, non-metallic compression rings, usually a PTFE composite, and frequently the proprietary Dixon product Rulon LD™. A story, much in favour of Rulon, circulates, although the source is not known. At the height of General Motors' involvement, seal performance deteriorated; after much valuable research effort had been diverted in search of a reason, it came to light that the purchasing department had been persuaded to buy a product manufactured in competition with Rulon LD (and presumably of the same, distinctive maroon colour!); when replacement seals were machined from the original Rulon, performance returned to normal.

Whatever the popular view to the contrary, the coefficient of friction, μ, of PTFE-based materials is *not* low, particularly at high sliding speed and low load, where it can reach 0.25 (Beldam-Crossley 1995). This is orders of magnitude greater than the value for hydrodynamic lubrication. Other things being equal, reducing leakage means increased interface force and thus increased friction loss. At high values of μ achieving the right balance is critical.

An interesting contrast arises with oil-free compressor technology, which favours the use of carbon- or graphite-based materials. There is a possible explanation for the different choice: some time after 1991 the specification of Rulon LD was changed to conform with international environmental and health and safety legislation. Ingredients had included cadmium sulphaselanide and barium sulphate (Beldam-Crossley 1991); the former came under suspicion as a potential health hazard and was replaced. Oil-free compressors are used to compress breathable gases. No doubt these are filtered, but the constitution of wear debris must be a concern and thus a factor in the choice of seal material. The UK manufacturer of Rulon has used iron oxide as the replacement, suggesting that the cadmium compound had been merely a colouring agent, and designated the result Rulon LR. However, performance in some applications has again fallen short of that of Rulon LD , and the oxide filler has been replaced (Greenwood 2000). If this explains why oil-free compressors tend not to use machined, PTFE-composite seals, it does not explain why Stirling engine technology ignores the availability of off-the-shelf – and thus relatively cheap – compression rings and penguin ring sets accurately moulded from carbon and graphite compounds.

7.4 Multi-cylinder rhombic engines

Other important collaborations picked up where the General Motors' program ended, some, no doubt, on the strength of the attractive features of the rhombic drive units. On the other hand, Philips were alone in continuing significant development effort on this configuration. They took the concept as far as a four-cylinder, in-line engine – the 4-235 shown in Fig. 7.4. In the process they uncovered perhaps the only technical disadvantage of a drive mechanism which otherwise answers all the kinematic and dynamic problems of the Stirling engine: it offers no economies when carried over to multi-cylinder, in-line form. Each cylinder still required a displacer, a power piston and its own combustion chamber, each combustion chamber calling in turn for an air pre-heater. Six connecting rods per cylinder unit is an excessive overhead.

The problem was elegantly mitigated in the 'flat-four' marine engine which halved the number of drive components by rigidly connecting opposing displacers, driving them from a common yoke, and dealing with opposing pistons in similar fashion. However, four independent combustion systems were still called for and specific power remained half that of a four-cylinder, 'double-acting' machine of the same swept volume. The height of the cylinders reflected the fact that they accommodated both piston and displacer. The inherent simplicity of the original invention had given way on all conceivable fronts to the opposite – to the ultimate in sophistication. Perhaps such a contradiction is self-destructive.

7.5 A widening of involvement

Discussions in 1964 paved the way for a development program with the German consortium MAN/MWM, and a licensing and 'know-how' agreement was in place between Philips and the Swedish partnership United Stirling (USS), established initially to develop the engine for use in mines. Later, Philips were to collaborate with the Ford Motor Co. of the United States on the development of a double-acting, swash-plate engine specifically for automobile propulsion. The USS venture resulted in an elegant, single-acting, opposed-piston engine in 'V'-configuration denoted the V-160 shown in Fig. 7.5. 160 is the volume in cm^3 swept in either of the cylinders (and *not* the swept volume or cubic displacement). All other United Stirling engines of any consequence have been of four-cylinder, double-acting type.

7.6 Back to thermodynamic design – via an anomaly

The multi-cylinder types have single combustion chambers, calling for the expansion exchanger of each cylinder to be one-quarter of the complete, symmetrical tube bank. The exchanger of the Ford 4-215 engine is representative. Individual quadrants are different in overall envelope from that of the symmetrical heater cage of the single-cylinder engine. The V-160 expansion exchanger too, differs in terms of external geometry from the Philips

Fig. 7.4 Philips' four-cylinder, rhombic drive bus engine. Courtesy of Philips Electronics BV

designs from which it ultimately derives.

There remains, however, a feature linking the exchanger systems to each other and to their origins. Figure 7.6 shows the expansion exchanger tubes of relatively large length and internal diameter compared with the shorter length and considerably smaller internal diameter of the tubes at the compression end. Internal diameter and length tend to be about one-third of the respective values at the expansion end. The feature will be found in all derivatives through to the Advenco (Advanced Engine Concept) of the late 1970s, and from limited graphical information available appears to be a feature of designs developed considerably later – and from an independent starting point (Carlsen 2000).

The differences in duct internal dimension and length can be loosely equated to corresponding differences in working fluid properties between the two exchangers: a factor of 3 in the densities, a factor of 1.5 between the coefficients of dynamic viscosity, and a variable factor between peak mass rates depending on the engine under consideration. But what of the regenerator

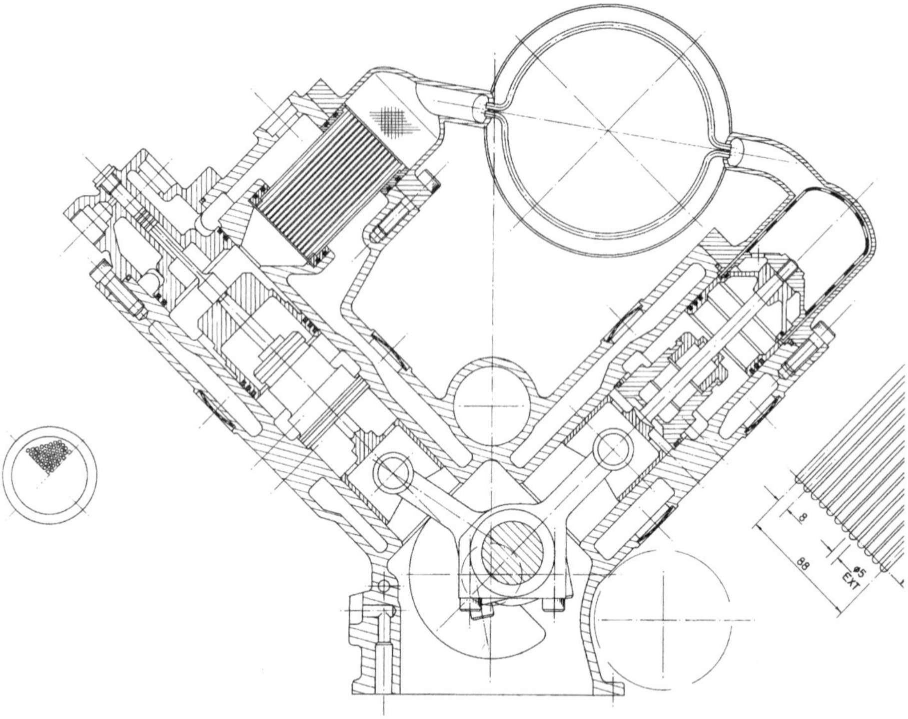

Fig. 7.5 Development of the United Stirling V-160 engine is now in the hands of SOLO of Germany who are incorporating it into a CHP package. After Organ (1997a) where the drawing is reproduced to a larger scale and is accompanied by a detailed specification

– the core component on which thermal efficiency and specific power crucially depend? It spans *precisely* the same range of temperature as the exchangers. Expansion and compression ends are subject, by definition, to the fluid conditions at outlet to the respective exchangers. Yet in all cases known to the writer, *a uniform mesh is used throughout*. (The regenerators of some miniature, reversed Stirling cycle coolers are an exception.)

The reader may not agree that something is necessarily wrong, but may be prepared to concede that something is not right! An improvement of 1 per cent in regenerator recovery ratio is widely claimed to be worth three times that percentage in thermal efficiency. Resolution of the anomaly must surely be worth 1 per cent, and the corresponding 3 per cent must surely justify the effort. Pursuing the investigation calls for a digression to continue the investigation of the 'regenerator problem'.

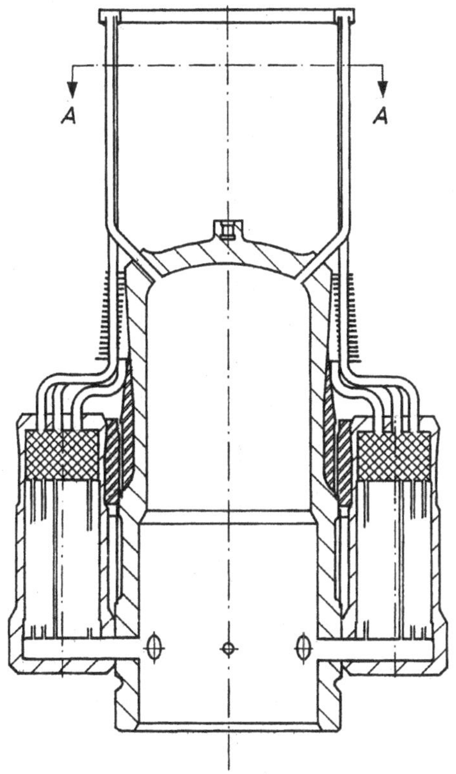

Fig. 7.6 Typical gas circuit illustrating characteristic geometric differences between expansion and compression exchangers. Reproduced with permission of Philips Electronics BV

8

The 'regenerator problem'

8.1 What regenerator problem?

The so-called 'regenerator problem' is a description, in mathematical form, of an idealized perception of regenerator operation. The description is emphatically *not* a faithful representation of actual conditions. On the other hand, it aims to capture the essence. The resulting stylized representation may be called the 'canonical' form, and the insight sought through its solution is the temperature–time history of fluid and matrix at all locations in both, and over as many cycles as it takes to achieve cyclic equilibrium.

A solution at this level may be expected to point the way to dealing with a problem description more closely resembling actual conditions. This is the justification for the considerable effort of cracking the canonical problem.

The original formulation addressed the operating conditions of the furnace air pre-heater. It may be re-expressed in Stirling engine parlance without changing it in any way. On this basis, Fig. 8.1a indicates a porous matrix contained within a duct of uniform cross-sectional area. It is assumed that:

(a) the fluid is incompressible, so that density, ρ, is uniform and constant;
(b) when fluid enters from the left it does so at uniform temperature, T_E; when it enters from the right it does so at uniform temperature T_C;
(c) the specific heat of the fluid, c [J/kg K} is constant and independent of temperature;
(d) there is no resistance to flow; pressure, p [Pa] is uniform and constant;
(e) local, instantaneous fluid temperature is T_g; that of the matrix is T_w;
(f) specific heat, c_w, and density, ρ_w, and of the parent matrix material are uniform and independent of temperature; volumetric porosity, ¶$_v$, is uniformly distributed;
(g) the thermal conductivity, k_w, of the matrix is zero in the flow direction and infinite perpendicular to it;
(h) local, instantaneous rate of heat exchange per unit area is proportional to a constant coefficient of convective heat transfer, h [W/m^2 K] and to local, instantaneous temperature difference, ΔT [K] = $T_w - T_g$;
(i) mass rate, m', and temperatures are uniform across a section perpendicular to flow direction;
(j) fluid mass rate is uniform between switchings; the uniform density

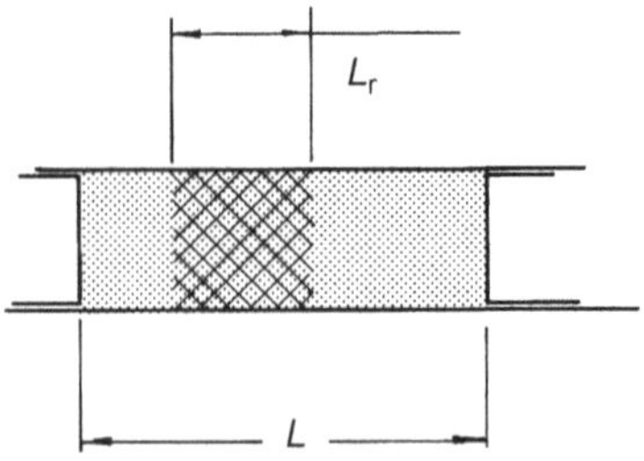

(a) Notation for regenerator operation

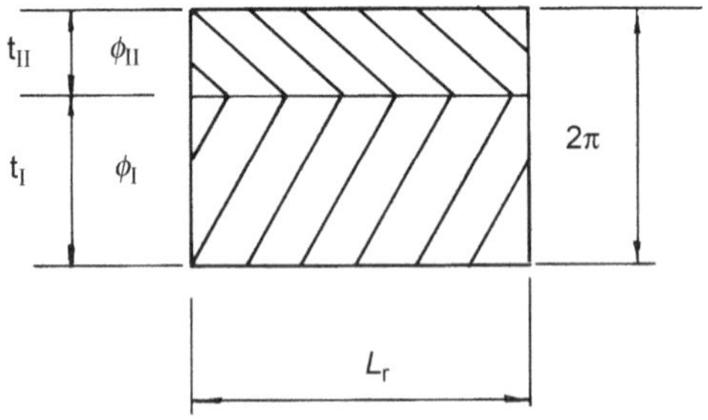

(b) Fluid particle paths for uniform mass rate and for asymmetrical blows

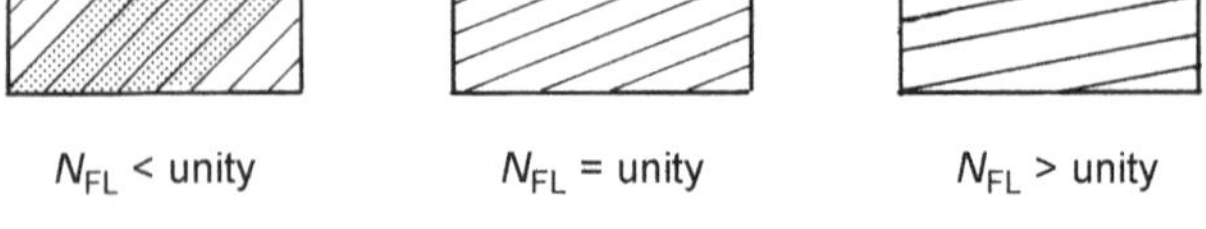

(c) Trajectory maps for varying flush ratio, N_{FL}

Fig. 8.1 Schematic representation of regenerator operation indicating particle motions for flush ratios of less than, equal to, and greater than unity

assumption means that velocity is uniform also;
(k) prior to the initial blow, matrix temperature is uniform at some value different from that of the incoming fluid.

The supposed fluid motion can be achieved if the duct is closed at opposite ends by pistons moving in synchronism. To this extent the problem, together with any available solutions, maps directly from the original context of the furnace air pre-heater to the Stirling engine. Further links are formed if initial volume between left-hand piston face and matrix is denoted V_E (for expansion space swept volume), and if the volume through which the right-hand piston expands is denoted V_C.

Rightward movement of the piston through volume V_E may also be thought of as the first left-to-right blow of a complete cycle occupying 2π radians of crankshaft rotation and involving a subsequent blow in the reverse direction. In the general case forward and reverse blows are asymmetrical (Fig. 8.1b) and of respective durations t_I and t_{II}. Presentation is slightly simplified by assuming symmetrical operation, which makes the initial blow of angular duration π radians.

The 'regenerator problem' or 'conjugate heat exchange problem' is to determine the temperatures $T_g(t,x)$ of the fluid and $T_w(t,x)$ of the matrix, as functions of time and location, from the start of the first blow, through subsequent reversals and through to cyclic equilibrium.

8.2 Early part-solutions

Schumann's solution of 1929 to the initial, left-to-right blow is exact for any degree of the initial flush, for any heat transfer coefficient and for any thermal capacity ratio. By the chronological point reached in this account the Schumann solutions had been used in the acquisition of heat transfer data for wire screens (Kays and London *op. cit.*). In principle, at least, these were now available for use in Stirling regenerator design. This, however, would call for their being 'fed back' into a model of the gas processes at cyclic equilibrium. It has been one of the remarkable features of the evolving discipline of Stirling cycle thermodynamics that simulation of the complete gas cycle (including, therefore, the heat exchange and flow friction processes in the regenerator) was claimed decades before the much simpler canonical problem had been properly solved!

This aspect of the regenerator problem, namely the analysis dealing with repeated flow reversals, had by this stage (1957) attracted Jakob's verdict as belonging *'the most difficult and involved that are encountered in engineering'*. Despite onslaughts by analysts of the calibre of Nusselt (e.g. 1927) the only indication as to how the temperature solutions might look at cyclic equilibrium had come from a compromise treatment by Hausen (1930). This depended on a simplification equivalent to the assumption of infinite flush ratio. The approach yielded values for recovery ratio, η_T, in function of thermal capacity ratio and *NTU*, and thence to the famous 'design' chart reproduced here as Fig. 8.2a. Possibly for the (unsatisfactory) reason that there was no alternative, the chart has for some years been promoted (e.g. Walker 1980, Kolin 1991) as a design aid for the Stirling regenerator. Four decades later, with flush ratio taken fully into account it becomes possible to assess the error. Comparison of Fig. 8.2b with the Hausen original confirms that the canonical problem yields a *minimum* recovery ratio of 50 per cent *even under adiabatic conditions*. The source (Organ 1997a) explains the apparent anomaly.

At the stage reached in this account digital computing power remained limited, relevant analytical solutions had not been acquired, and it can be claimed with confidence that no Stirling engine so far designed has benefited from any validated insights into regenerator transient thermal response.

8.3 The makings of cycle analysis

Prospects were shortly to change, however. Some years previously, and possibly as their central design resource, Philips had rediscovered the now ubiquitous analytical treatment of the Stirling cycle by Schmidt (1871). On the basis of some drastic assumptions about the gas processes, this furnishes a prediction of pressure variation with crank angle, an indicator diagram, and an estimate of cycle work. The Philips reworking (Rinia and duPré 1946) was an advance over the original treatment for being developed in terms of *dimensionless variables* throughout.

Those who have grasped the concept of temperature ratio, volume ratio – or even of bore-to-stroke ratio – need not fear dimensionless variables. All properly formulated equations are *dimensionally in balance*: the dimensions [Mass, Length, Time, Kelvins (temperature)] of the left-hand side of the equality sign (=) necessarily balance those on the right. Any term common to the right side may be extracted and put into the common denominator of the left without affecting the balance – or the meaning of the equation.

Judicious transposition of this sort results in *dimensionless groupings* of variables. Temperature ratio, $N_T = T_E/T_C$ is a simple example, and illustrates a crucially important point: absorbing all occurrences of T_E and T_C into the single symbol N_T *reduces the number of parameters of the problem by one!* More generally, an important theorem of equations (White 1979) is that by normalizing (making dimensionless) an equation having *n* parameters, it can be reduced to an equivalent dimensionless equation having *up to four parameters fewer than the original*. For the maximum reduction of four to be achievable,

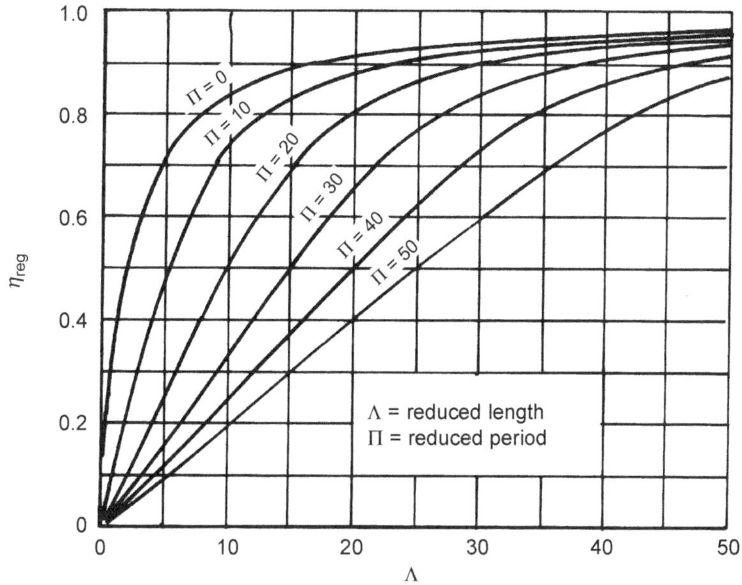

(a) Hausen's celebrated temperature recovery ratio curves. Reproduced with permission of Springer-Verlag

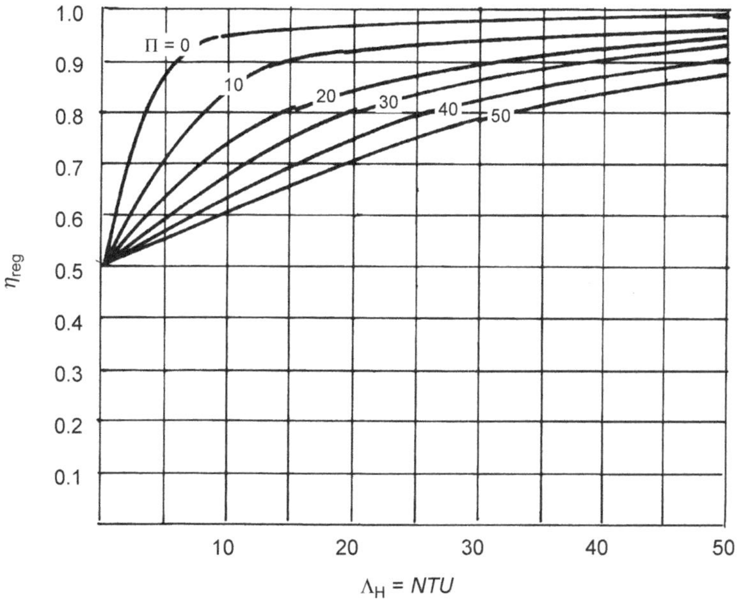

(b) Design curves from the identical problem with flush ratio taken into account

Fig. 8.2 Comparison of solutions to the 'canonical' regenerator problem (a) without and (b) with account of the flush phase

the original parameters must embrace all four of the dimensions M, L, T, and K, but *any* reduction may be considered a benefit. The Philips' treatment made the cycle analysis a function of four parameters – N_T, thermodynamic volume ratio, and phase angle, κ and α, together with a 'dead space parameter' ν. Within the limitations of Schmidt's original rationalizations, the reformulation now applied to *any* working fluid provided only that it behaved as an ideal gas ($p/\rho = RT$).

However, it fell to Finkelstein (1960a) to draw out the full potential of the dimensionless form. His virtuoso analytical exposition applies with indifference to *any* configuration of engine – alpha, beta or gamma, and to any choice of reference charge pressure, p_{ref}, p_{min}, p_{max}, or p_{mean}. With hindsight, the *coup de grâce* lay in recognizing that the dimensionless formulation led to dimensionless solutions – not only to the 'obvious' ideal efficiency, $\eta_{Schmidt}$ = 1 - $1/N_T$ but to *specific cycle work* also. Four alternative forms were identified, one based on total fluid mass, M, and defined as $W/(MRT_C)$ or, equally, to *power*/$(MRT_C f)$. The other expressions were based on reference pressure and volume, and can be written *power*/$(p_{ref}V_{sw}f)$, in which p_{ref} may be p_{min}, p_{max}, or p_{mean} depending on how the designer sees the problem of charging the engine with working fluid.

The concept of specific cycle work opens up the prospect of *optimization*: there is little meaning to the notion of a 'best' work, W or power, both of which may be increased without limit by merely increasing size or pressure (or both). In contrast, peak *specific* cycle work is the maximum that can be achieved from the available mass, M, of working fluid, or, alternatively, from given swept volume, V_{sw}, and charge pressure, p_{ref}. Not content to leave us with the prospect, Finkelstein carried out the optimization process – and not with the aid of a computer, but in *tour de force* analytical style.

The ultimate value to thermodynamic design of specific cycle work is that the numerical value is *independent of size and charge pressure*. In other words, *all* 'Schmidt' engines having the same respective N_T, κ, α, and ν also have the same specific cycle work. A value can also be calculated for a real engine. Even for modern designs, the result falls short of the corresponding Schmidt ideal. On the other hand, for a large number of well-designed (or well-developed) modern engines operating with common temperature ratio, $N_T \approx 3.0$, the scatter of numerical values is minimal, most lying close to 0.15.

$$\frac{power}{p_{ref}V_{sw}f} \approx 0.15 \qquad (8.1)$$

Walker (1980) has proposed the label 'Beale number', N_B, for the ratio at the left-hand side. It is, however, merely the experimental counterpart of Finkelstein's definition of two decades previous, for which the symbol N_{TF}, for Finkelstein number, will be used here in recognition of the origin and significance. (The more obvious N_F is already in use for Fourier modulus.)

The 'optimum' Stirling engine determined by analysis can be no better than the assumptions of the underlying mathematical model – in this case, those of Schmidt's original formulation. Prominent among these had been that

(a) heat exchange between gas and adjacent wall is instantaneous and does not involve a temperature difference; and
(b) fluid flow occurs with zero pressure gradient, i.e. without viscous losses.

The gas processes of a Schmidt engine are independent of cyclic speed. The value of N_{TF} given by the Schmidt algebra is thus independent of r/min, meaning in turn that power increases linearly with speed, and without limit. This is in conflict with reality.

In response, Finkelstein formulated his *Generalized Thermodynamic Analysis* (1960b) in which instantaneous heat exchange in the variable-volume spaces was not infinite, as in the Schmidt model, but *limited* to the rate permitted by contact area, temperature difference, and coefficient of convective heat transfer, h. Computing facilities at the time amounted to an electromechanical desk machine, and calculations corresponding to a single revolution of the crankshaft occupied about 6 weeks' work, warranting, in Finkelstein's own words *'the use of an electronic computer'*! The equivalent computation, slightly modified (Organ 2000b), takes milliseconds on a modern PC., but it is significant that the original algebra of the variable-volume spaces has never been improved upon.

8.4 The advent of computer simulation

As well as leading to more realistic numerical estimates for power and efficiency, the generalized approach lent insights into the local gas processes: dimensionless formulation led to the variables σ and σ', respectively dimensionless mass, m/M, and dimensionless mass rate $\partial(m/M)\partial\phi$. Dealing in terms of mass *fraction* kept the treatment *independent of any particular working fluid*, as for the Schmidt analysis (one-tenth of the total mass inventory, M, of helium takes the same numerical value as one-tenth of M when M is air).

The *Generalized Thermodynamic Analysis* represented the first published computer simulation of the Stirling cycle. Its importance for air engines lay in the *parameters* whose numerical values determined the performance prediction. To the Schmidt parameters of N_T, κ, α, and υ were now added *specific heat ratio*, γ, and an *NTU* for each working space – NTU_e and NTU_c. A Finkelstein engine with helium (γ = 1.66) was now bound to perform differently from the identical machine charged with air or hydrogen (γ= 1.4 in both cases). On the other hand, air and hydrogen would, to this extent, offer the same efficiency and specific output. For all things to be equal, however, the *NTU* have to be maintained as the working fluid is changed. The definition of Finkelstein's *NTU* was

$$NTU = hA/\omega Mc_p \tag{8.2}$$

in which h [W/m² K] is the coefficient of convective heat transfer, A [m²] is exposed cylinder area, ω [1/s] is angular speed, M [kg] is the total mass of working fluid and c_p [J/kg K] is specific heat at constant pressure. Individually, these variables take different values for different gases. However, it is the numerical value when *combined* into *NTU* which determines the cyclic processes of the Finkelstein engine. Thus, there emerges for the first time the prospect of an air-charged engine redesigned to have the same gas processes, and thus the same cycle work, as one charged with hydrogen.

8.5 A first fluid particle trajectory map

It would take three decades for the implications for thermodynamic design to be fully recognized. The delay might have been longer but for a further, complementary insight achieved by another worker. A year after Finkelstein, Walker (1961) published an account of a largely experimental study of a Philips' Stirling cycle refrigerator. Almost as an aside it featured a *fluid particle trajectory map* [not quite the prototype of such maps, since Schöttler (1881) had plotted the approximate path of a specimen particle in his study of the Rider engine]. This is the gas circuit 'straightened out' into equivalent, opposed-piston layout. The vertical axis covers the 360° of a crankshaft revolution. The horizontal axis is proportional to the path length through the exchanger system with outer-dead-centre of the left (expansion) piston face as datum. The curved lines mark the computed paths of selected working fluid particles over a complete cycle.

The analysis underlying the map obviously embodies simplifications: Walker used the Schmidt analysis and, like Finkelstein, calculated the particle positions by hand *'at 5 or 10° increments of crankshaft rotation'* (Walker, 1999). This, however, reflected the computational facilities of the era rather than any inherent limitation of the concept.

Applied to a well-designed engine with modest flow losses (small pressure gradients) and adequate heat transfer (small differences locally between fluid and adjacent solid surface) it is unlikely to be grossly misleading. The mass of working fluid between successive particle tracks may be chosen to be constant, in which case counting particle tracks yields a ready estimate of flush ratio – the amount of working fluid which passes through the regenerator as a fraction of the resident mass. Thus it was verified for the first time that the flush ratio of the typical Stirling engine has a *very small value* by comparison with that of the furnace air pre-heater, usually between 0.5 and 3.

The trajectory map is set to assume such importance that the original is reproduced as Fig. 8.3. This significance stems from the facts that:

(a) like the *Generalized Thermodynamic Analysis* it is independent of working fluid,
(b) the extent of the flush process is shown at a glance: here for the first time was a compelling graphical representation of conditions in a Stirling machine, showing that, for the case under study, *no single working fluid particle passed completely through the regenerator.*

The established approach to regenerator analysis had been shown to be fundamentally inconsistent with conditions in the Stirling engine. Four decades would pass before Stirling engine and regenerator theory were fully reconciled, but a crucial process has begun: theoretical study was finally yielding *insight* rather than numbers. Moreover, pictorial display makes insight *communicable.*

8.6 Lateral thinking

An account of developments to the early 1960s would be incomplete without reference to a unique practical development – the *internally focusing Stirling engine* (Finkelstein and Eibling, 1961). The best explanation of its significance is in terms of the essential difference between the internal combustion engine and closed-cycle engines in general. In the former, the heat release is internal

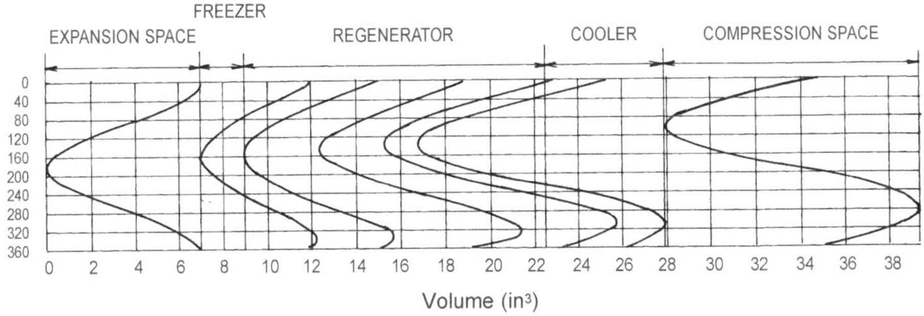

Fig. 8.3 The prototype of all fluid particle trajectory maps. After Walker (1961) with permission of the Institution of Mechanical Engineers

to the cylinder, is in the form of a chemical reaction between fuel and air, and is virtually instantaneous; in the latter, the thermal conduction through metal walls places a limit on the rate of energy input. The new solar air engine replaced the cylinder head by a quartz window (Fig. 8.4). Sunlight concentrated by a lens or mirror passed through the window and heated a porous absorber – essentially an extension of the regenerator made of refractory material.

To assess how far it went towards eliminating the thermal conduction bottleneck it would be necessary to know the transmission characteristics of the quartz window and other candidate materials. Despite the advantageous compression ratio allowed by elimination of heat exchanger tubes and associated dead space, achievement of useful power output would call for pressurization. The invention was shamelessly imitated (Organ, 1962), but never, as far as is known, exploited as it stood or developed further.

Sufficient concepts are now in place to make possible another view of the Stirling engine gas processes which have a bearing on the evolving principles of thermodynamic design.

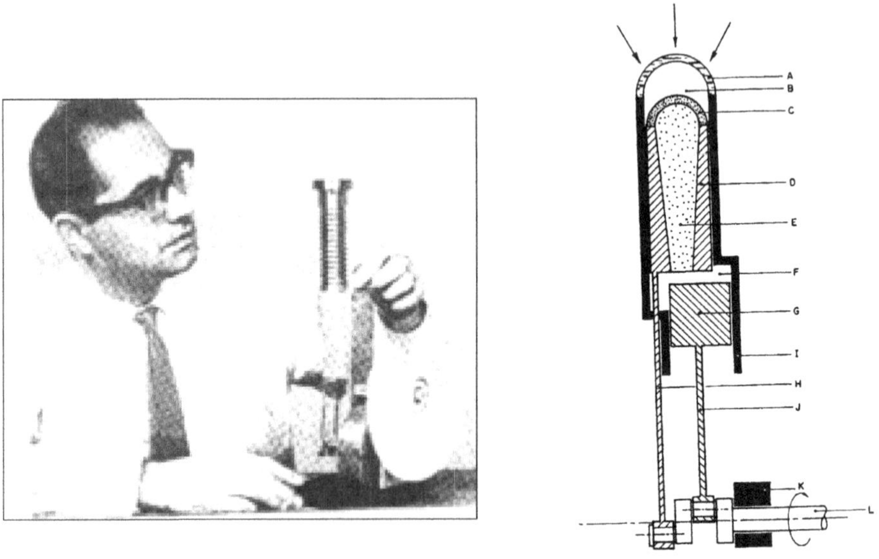

Fig. 8.4 Finkelstein's internally focusing solar engine – the first and only internally heated Stirling cycle engine
Reproduced from Finkelstein and Eibling (1961) with permission of the American Society of Mechanical Engineers (ASME)

8.7 Air versus helium versus hydrogen

There can surely be few instances where the fortunes of a branch of engineering science have been so influenced by the appearance of a single diagram. This, however, may be an understatement in respect of the air engine and of Fig. 8.5. The latter appeared in 1965 and summarizes computer-simulated performance of a 'beta'-configuration (displacer-type) Stirling engine of about 50 b.h.p. per cylinder. The curves compare the performance of designs optimized for efficiency on three working fluids. The apparent superiority of hydrogen and helium over air has been accepted without question ever since, and has almost certainly deflected attention from air-charged (or nitrogen-charged) designs. However, there is an alternative, and very different, point of view.

The physical processes by which heat is converted to work in the Stirling engine are forced convective heat transfer and the associated viscous dissipation, or flow friction. It will be recalled (from Chapter 6) that the heat transfer data for all the common gases collapse onto common respective curves when plotted in terms of N_{st} (Stanton number) against N_{re} (Reynolds number).

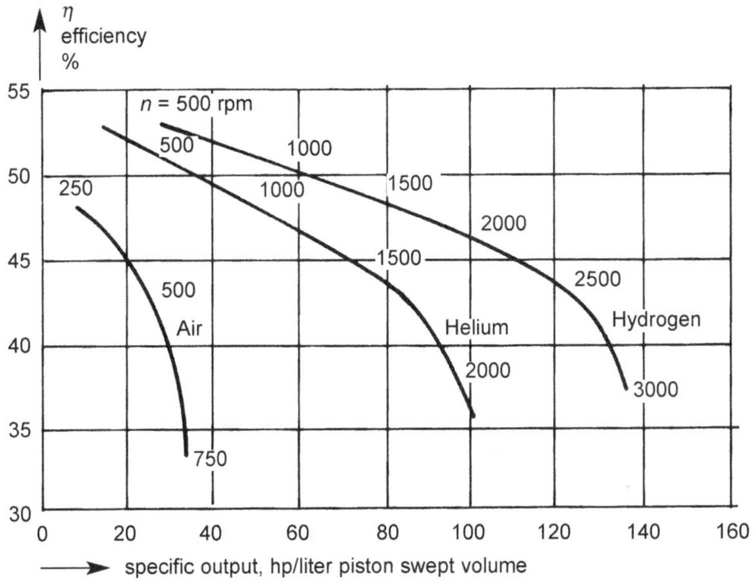

Fig. 8.5 A much-quoted diagram depicting the computer-simulated effect of working fluid on the thermal efficiency of the 'optimum' Stirling engine. Reproduced with permission of Philips Electronics BV

The same is true for flow friction data plotted as friction factor, C_f. Reynolds number, N_{re} is defined as $4\rho \underline{u} r_h/\mu$, so that a specific numerical value is determined not merely by coefficient of dynamic viscosity, μ, but by particle speed and thus by engine r/min. It is also influenced by density, and thus by a *combination* of charge pressure, *p*, with the fluid characteristic, *R*. Stanton number, $N_{st} = h/\rho \underline{u} c_p$ likewise contains $\underline{u}$, which is independent of working fluid. Moreover, conversion of N_{st} to the compound parameter *NTU* which determines the exchanger temperature profile involves multiplication by the geometric parameter L/r_h whose value is *independent of any working fluid*. On this basis, any value of N_{re} and thus any values of *NTU* and C_f achieved with hydrogen as the working fluid may, in principle, be achieved with air by appropriate adjustment to the various hydraulic radii, r_h, and/or exchanger and regenerator lengths, *L*, and/or rotational speed and/or charge pressure. Such adjustments promise to replicate with air the distributions of temperature and pressure achieved with the 'superior' gas. In addition two engines with gas processes that are similar to this extent may be expected to have similar specific power and thermal efficiency.

A large body of evidence has since been accumulated in support of these 'Similarity' principles. Some derives from reinterpretation of published specifications and performance data, some from specially devised laboratory experiments. The author's 1997 text documents both, and spells out the arithmetic of design by thermodynamic similarity. Chapter 10 will take the present account further. In the meantime, a previously unpublicized observation is relevant. This arose during the development of a reversed Stirling cycle cryo-cooler designated the NAX-106. Some details of construction are unclasssified (Organ 1997a), but quantitative performance data are the property of the Ministry of Defence. It is nevertheless permissible to disclose that the rated working fluid was helium – and that tests were carried out with alternatives. Over a wide range of charge pressures, initial cool-down rates with nitrogen markedly surpassed those achieved with the fluid for which the cooler had been 'designed'. ('Initial' rates because nitrogen is not a suitable working fluid for cooling down to the rated operating temperature of 77 K.) Among other things, the finding would support a serious evaluation of nitrogen as working fluid for CFC-free domestic refrigeration.

Paradoxically, the strongest argument in favour of Similarity comes from the sceptics, whose counter-argument goes as follows. Take a hydrogen-charged engine (the 'prototype') and scale the gas circuit for comparable performance with air at the original charge pressure and r/min. Assuming the performance expectation is realized, recharge the new engine (the 'derivative') with H_2 to the original pressure and operate at the original r/min. By the principle of the alleged superiority of the lighter gas (c.f. Fig. 8.5) performance must improve over that of the air-charged derivative – and thus, by definition, over that of the prototype.

A practical experiment based on the hypothesis promises to be of greater interest – regardless of outcome – than any so far conducted in support of Stirling engine design. Suppose first of all that the gas circuit design of the prototype is an absolute 'optimum' at rated r/min and charge pressure. This means, *by definition*, that *any change* will result in a deterioration in thermal efficiency, power, or both. With exchangers and regenerator respecified for air, reversion to H_2-charging *must necessarily* result in performance deterioration.

Now suppose that the prototype is *not* optimum, as is likely in practice. It is now conceptually possible that the performance of the derivative, when recharged with H_2, will exceed that of the prototype. In this case, the scaling exercise has, in a round-about way, led to a welcome design enhancement.

The former outcome lends further support to Similarity principles for gas circuit design. The latter shows the way towards the optimum H_2-charged engine. Who is going to complain about either?

9

Two decades of optimism

9.1 Summary

For the understanding of cycle thermodynamics and of the technology of construction, the period from the mid-1960s to the mid-1980s was essentially one of consolidation. Rapid advances in computing capacity encouraged the development of ambitious simulations of the gas process cycle. With the notable exception of a contribution by Miyabe et al. (1982), however, study of the regenerator continued in isolation and without impact on Stirling engine design.

The era saw a number of new embodiments. The free-piston variant was invented (Finkelstein 1963a, b, 1964). West (1970) proposed the liquid-piston or Fluidyne engine. Bradley (1974) demonstrated an engine operating from a low-temperature source, possibly the first of a type to become known as the 'Low ΔT' genre.

9.2 The free-piston engine

It is possible to dispense with the displacer linkage, driving instead by the variable difference in pressure between the gas circuit and the buffer (or crank-case). The result is the 'Ringbom' engine. The pressure difference acts on the displacer drive rod; due to its inertia, the displacer responds with a motion out-of-phase with the applied force.

The view (Urieli and Berchowitz 1984) that the free-piston type originated with William Beale might be disputed, but 1964 saw the start of a long-term commitment by him to the type. His independent approach may have been the result of noting that the functioning of the Ringbom engine does not depend on the piston itself being kinematically driven: a piston suspended by a spring will oscillate with more or less simple-harmonic motion. Adding a spring-mounted Ringbom displacer gives a prime mover with no rotating parts.

Figure 9.1 is a cross-section through a representative Beale engine. The 'thermodynamic end' has much in common with that of the Philips MP1002CA: an annular regenerator and heat exchangers consisting of axial slots. The springs function on the principle of the loud-speaker suspension, providing rigid radial location as well as the axial stiffness [N/m] called for by operating frequency and the mass of the reciprocating member. The combination of suspension and radial location gives the free-piston engine an important edge over the traditional embodiment: side loads on seals and bearing surfaces are negligible.

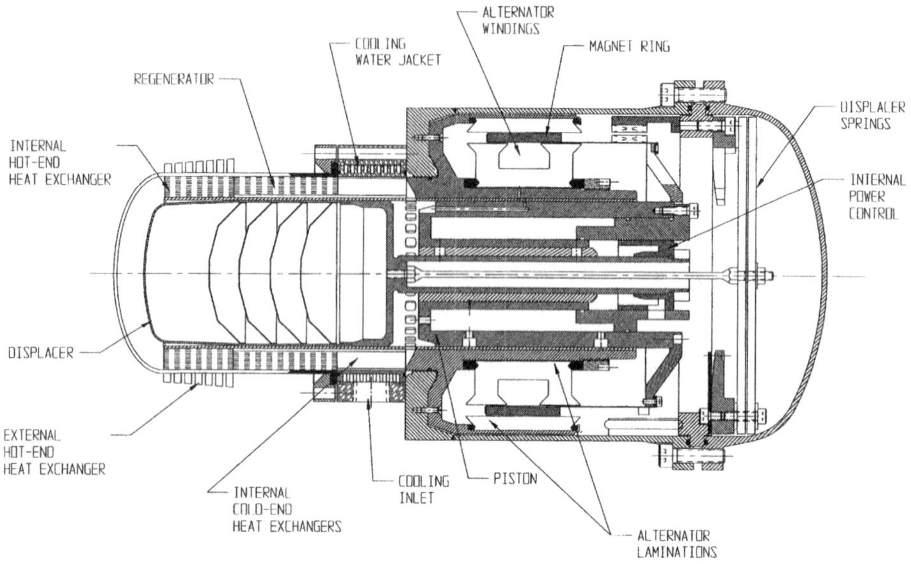

Fig. 9.1 A Sunpower free-piston Stirling engine. Illustration supplied by, and reproduced with permission of, Professor William T. Beale

9.3 The fluidyne

The liquid-piston or 'Fluidyne' engine was invented at the UK Harwell Laboratories. In this delightfully simple embodiment the gas circuit joins the tops of two tubes containing a liquid; water is suitable. A tuning or output tube connects to the limb of the original U-tube close to the expansion space end. The inventor describes operation with reference to a figure reproduced here as Fig. 9.2.

Oscillation of the liquid in the U-tube unaccompanied by any movement of the liquid in the output tube would represent pure displacement, with the working gas (usually air) being moved between hot and cold spaces without overall volume change. On the other hand, movement of liquid in the output line does result in a change of volume – like the power piston in a gamma Stirling engine.

When the liquid in the open end of the output tube is moving downward, the gas in the working space is being compressed. When the fluid is moving upward, the gas is being expanded. For the machine to produce work as a Stirling engine, expansion must take place with most of the gas at the hot end

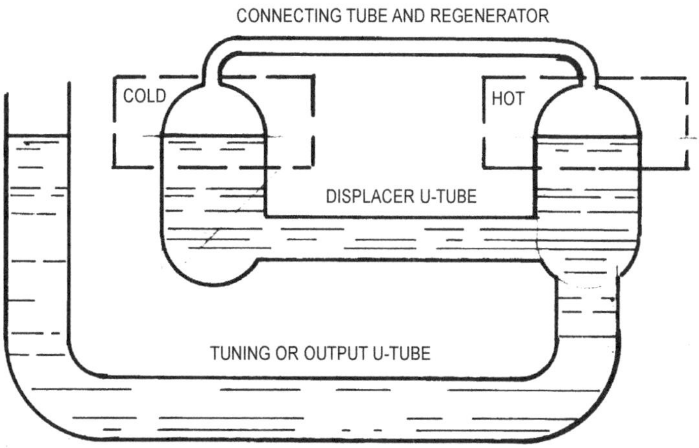

Fig. 9.2 Principle of the Fluidyne engine

of the displacer, i.e. with the liquid level lower at the hot end of the displacer than at the cold end. Compression must take place with most of the gas in the cold space.

In order for this to happen, the junction between displacer and output tubes must be . . . . located closer to the hot than to the cold end of the displacer. The two sections of the displacer liquid column, between the junction and the free surface, are subject to the same pressure difference (if pressure drop across the regenerator is ignored). If the hot section of the column is the shorter, as in the figure, it will respond to the pressure more readily than the greater mass of liquid in the cold section, and its movement will be more advanced in phase.

The phase advance will be recognized as that required to cause operation as a Stirling engine. The principle can be elegantly demonstrated in miniature in a system of transparent Pyrex tubes. A large-scale version built by the Metal Box Co. of India Ltd is reported (West *op. cit.*) to have pumped 4000 gal/h against a head of 12 ft. At first sight a technology compatible with the resources of developing countries, the Fluidyne appears not to have been exploited. It does not lend itself to pressurization above 1 atm, and operating frequency is inherently low. The pumping power of the Metal Box engine converts to 165 W for 4.4 kW input, or a thermal efficiency of 3.75 per cent. A regular energy supply equivalent to 4.4 kW from local resources is not necessarily an easy proposition, and the provision, including stoking, could readily account for manpower equivalent to engine output (165 W).

9.4 The Low ΔT variant

Bradley Engineering was a one-man business based largely on the sale of batch-produced, die-cast cylinder sets (Fig. 9.3), accessories and drawings for model engineers. The displacer cylinder units had three distinctive features: (a) large ratio of diameter to stroke, (b) cast-in 'pin-fins' protruding both internally and externally from the plane end faces to enhance heat transfer, and (c) an 'inside-out' displacer design best described as a pair of flanged washers back-to-back, the flanges facing outwards in opposite directions. Like the cylinder unit, the displacer had a large ratio of diameter to length.

One application of the cylinder units was a rocking beam engine. Under contract to the University of Calgary the design (Fig. 9.3) was scaled up and the 'pin-fin' concept taken to its logical limit by brazing a close-packed array of small-diameter copper rods into holes drilled through the cylinder end plates. The rods protruded approximately 10 mm either side, affording large surface area together with a high conductivity path through the high-strength end plates. Intended for high-temperature operation from the heat of a gas burner the engine would run unloaded with the expansion end warm rather than hot.

Andrew Bradley died prematurely, most of his engines and stock of cylinder sets going to a collector from the United States. He might otherwise have been known as the father of the Low ΔT engine. As it is, the debut of the type is set by Kolin (1991) at 1983. Kolin's substantial contribution to Low ΔT technology has been to appreciate that the low temperatures permit the use of materials having properties superior for the purpose to those of metals. Thus, in the case of the displacer, reciprocating mass, and thermal shorting may both be minimized by use of polystyrene. Likewise, thermal shorting along the displacer cylinder walls may be all but eliminated by use of a suitable plastic.

The proprietary Low ΔT engine shown in Fig. 9.4 has a diaphragm compression piston linked kinematically to the displacer by a Ross drive (Ross 1983). Other variants drive the displacer by the Ringbom principle. Senft (1993) has made notable contributions to the understanding of the dynamics of the Ringbom variant. Over a period, design of the type became competitive, the goal being no-load operation from the minimum temperature difference. It is due largely to the combined efforts of Kolin and Senft that it has been shown possible to design for operation between temperatures differing by a mere 1 °C. Kolin's (1991) graph, reproduced here as Fig, 9.5, overlooks Bradley's contribution, but shows Senft's remarkable achievement of a Stirling-type heat engine operating between source and sink a mere 0.5 °C apart!

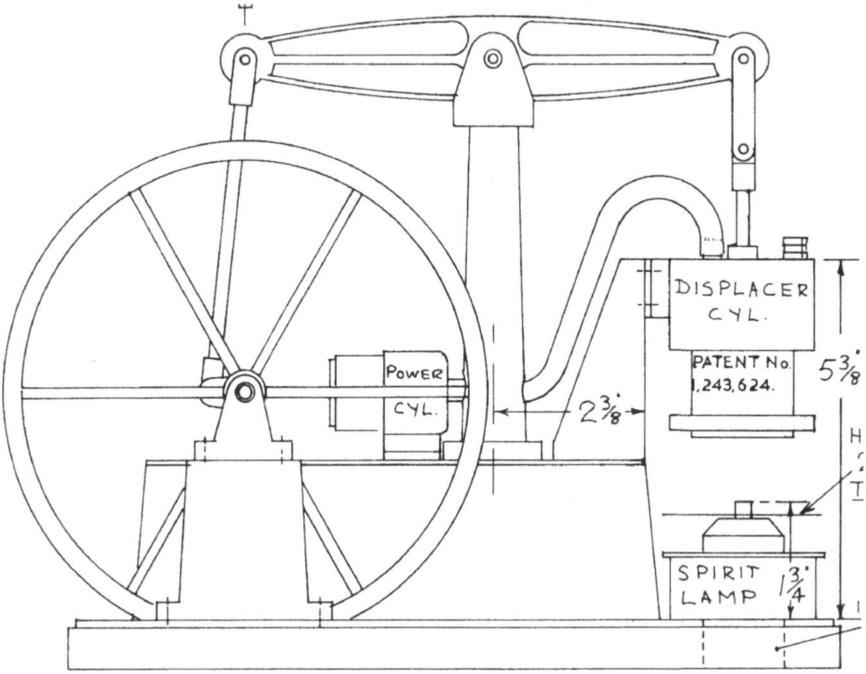

Fig. 9.3 Bradley cylinder unit and modular engine featuring 'inside-out' displacer. Scaled up for high-temperature operation (~ 600 °C) the rocking beam version also ran unloaded as a 'Low ΔT' engine (i.e. at very small temperature differences). Drs Rod Fauvel and G. Walker kindly tracked down and provided the drawing. Both illustrations are by Bradley himself

Fig. 9.4 Proprietary Low ΔT engine manufactured by the New Machine Co., Kirkland, Washington, USA

Low ΔT engines are characterized by small values of kinematic volume ratio, λ. The large-diameter displacer cylinder is an ideal receiver for solar radiation, and there is immediately an alternative to the high-temperature solar engine discussed earlier.

There is also an interesting thermodynamic insight: in a conventional, high-temperature Stirling engine a design goal is to maximize compression ratio. In the Low ΔT type the 'adiabatic' component of temperature swing of a high compression ratio would readily exceed the cycle temperature limits, inhibiting the required heat flows. Looking again at the relative volumes swept in the Low ΔT type, it can be seen that thermodynamic phase angle, α, is determined essentially by displacer movement alone, and is thus close to 180°. Assuming a value for kinematic displacement ratio, λ, of 0.1 permits solution of the equivalence equations [equations (6.1)–(6.5)] for thermodynamic volume ratio, κ; this turns out to be close to unity. The values may be compared with those from design charts produced by computer optimization of the 'adiabatic' cycle – for example, by those of Organ (1992a, pp. 351–356). The relevant value of temperature ratio, N_T, when ΔT is small is essentially unity. It will be seen that κ converges on a value of unity for all cases examined, while phase angle, α tends to a value of π (i.e. to 180°) in all cases also. The counterpart Schmidt charts (pp. 346–350) give a different result because the treatment completely suppresses the adiabatic component of temperature swing.

The Low ΔT engine produces no net output power, and to this extent remains a toy. On the other hand, its impact in that role has drawn attention to the

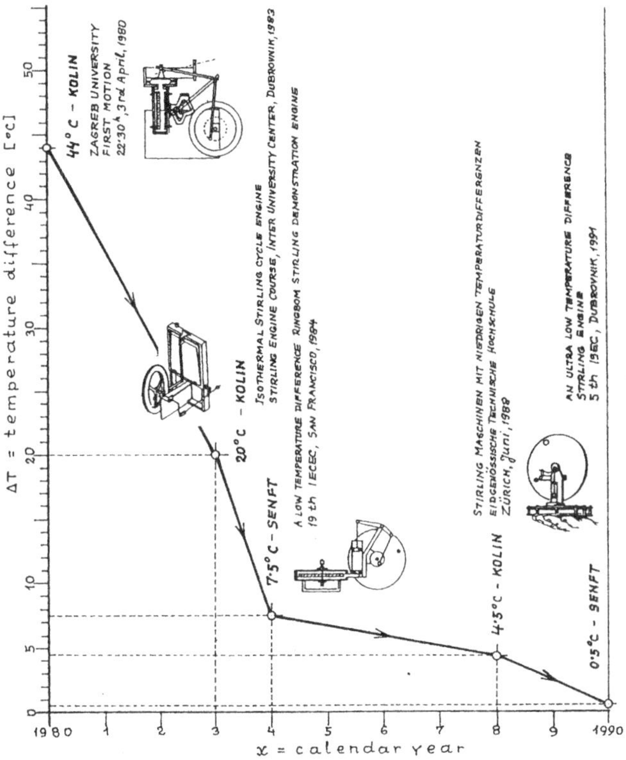

Fig. 9.5 Plot of progress of a competitive effort to produce an engine to operate unloaded from the minimum temperature differential, ΔT. Reproduced from Kolin (1991) with permission

possibilities for *intermediate-temperature* variants constructed from unsophisticated materials. James (1999) is developing engines for waste heat recovery along these lines.

9.5 The era of the computer

In view of repeated claims for its contribution to engine design, it is a pity not to be able to discuss Philips' computer simulation work. Hargreaves' definitive account (*op. cit.*) serves only to confirm that disclosure was never intended. Reading between the lines is not an exact science – and probably not a science at all – but is the only recourse.

The ten-volume report (Cuttil *et al.* 1962) by the Alison Laboratory on the

PD-46 satellite engine development program describes a computer model of the cycle intended to supersede the Philips' code which '... *assumed that the pressure was identical throughout the system and that the mass transfer within the system was dependent only on the displacement of the pistons and the temperature changes within the system'*. When attention turned from space to terra firma, Tew *et al.* (1978) felt the need to develop a computer simulation from scratch to support development of the GPU-3 engine. With Philips providing the 'know-how' under both licensing agreements the inference is that the gap between cycle modelling and thermodynamic design had not been bridged to the satisfaction of the licensees.

The only specific indication as to how the task was being approached comes from an account by Feurer (1973) of computer simulation work carried out while the MAN/MWM consortium was a Philips' licensee. This was strongly indicative of a key simplifying assumption, namely, that the various processes – volume variations, pressure variations, etc. – could be assumed to be 'simple harmonic', i.e. to be capable of description by a sine (or cosine) function. The gas processes thus predicted were used to estimate the imperfections or 'losses' of the corresponding 'real' effects of flow friction, limited heat transfer, etc. The inference is consistent with Meijer's comprehensively documented account (1960) of mechanical aspects of engine design. This focuses on the kinematics of the rhombic drive, analysing the complex motion in terms of an infinite series of individual simple harmonic components. The approach would dovetail with the linear harmonic approach to simulation, since the first harmonic of the rhombic drive motion becomes the (simple) harmonic volume variation.

A paper which appeared in 1978 by Fokker and van Eckelen raises expectation of disclosure, but turns out to be a rehearsal of the universal principles involved in acquiring numerical solutions to the differential equations of one-dimensional fluid flow and heat transfer. Perversely, any solutions of interest to Stirling engine thermodynamic design are withheld.

Consecutive papers by Schalkwijk (1958, 1960) give an initial expectation of affording insight into regenerator design. On closer examination the treatment is found to ignore the underlying physical processes in favour of a purely algebraic manipulation of Hausen's result of decades earlier. Since the latter suppressed the flush ratio, the contribution to insight was nil.

Whether engine design within Philips benefited from cycle analysis or computer simulation may never be revealed. The outside world gained inspiration from a succession of sophisticated engines operating on helium and hydrogen – but not a single guideline for gas circuit design. There was accordingly *still* no means to a scientific evaluation of prospects for the air-charged variant.

Academic scientists working independently on cycle thermodynamics and simulation took the opposite approach to disclosure, with full details of underlying physical model, numerical integration scheme, specimen results – and, in some cases (e.g. Urieli 1977, Urieli and Berchowitz 1984) publishing the computer code in full. Two features characterize all simulation work of the era.

(1) The gas circuit is broken down for the purposes of numerical processes into a succession of adjacent 'cells' and associated 'nodes' and the defining equations solved for each cell in succession (say, left-to-right) or simultaneously in an automated process equivalent to 'relaxation'. This forces gas particles and pressure information to propagate at a common speed, namely, that of the integration process, $\Delta x/\Delta t$, where Δx is the (arbitrarily) chosen length of the cell and Δt is the (equally arbitrarily) chosen increment of time. In reality, pressure information propagates at acoustic speed, a, ± particle speed, u, and particles at their own, much slower velocity u.

(2) The algebraic formulation was generally in *absolute variables* – r/min (or angular speed, ω [rad/s]), absolute pressure, p [Pa], etc. Even in more forward-thinking formulations such as that of Urieli, resulting output therefore tended to have more to say about a *specific engine design* than about Stirling engines in general.

The simulation era created a new resource. However, either you had a simulation and could thus claim to be able to design a gas circuit from scratch, or you hadn't – and couldn't. If there was any *communicable insight* to be distilled from the number crunching, no one organized it for publication. Aside from independent support from Urieli (*op. cit.*) for Philips' earlier finding that an engine designed for hydrogen performs differently on air, there was still no generally accessible means of designing specifically for one or the other. A design manual by Martini (1978) amounted to a stencil for laborious, hand-calculation: individual loss components – thermal short, heat transferred by the shuttling displacer, regenerator pumping loss, etc. gave numerical values to be subtracted from power and efficiency of an 'ideal', lossless cycle. A revision of the design called for the calculation to be repeated.

9.6 Further advances by Philips

To this point the air-charged variant remains a poor relation of the hot-gas engine family. It comes as no surprise that work of practical development proceeded largely around hydrogen and helium, and moreover with engines of multi-cylinder, double-acting layout.

One of the most ambitious ventures was the development of the Philips 4-65 DA double-acting, swash-plate engine into the 4-215 DA hydrogen-charged version for evaluation in a Ford Torino automobile. Figure 9.6 shows one of five prototypes ready for installation. Hargreaves (*op. cit.*) gives an absorbing and comprehensive account of the mechanical engineering problems and solutions, but no indication as to how the gas circuit design was arrived at. Performance and weight were close to expectation, but problems with combustion, power control, and with the reliability of the roll-sock seal contributed to a decision by Ford in about 1978 to discontinue development.

The problems with this complex engine were apparently not unexpected: in parallel with the 4-215 DA development the Advenco concept was taking shape. The name is an abbreviation for Advance Engine Concept, and put the emphasis on pragmatic solutions to the issues of power control, gas retention, and cost. Hargreaves details the approach to the problem of heater materials and fabrication. Power control was tackled by varying piston stroke rather than gas pressure. Figure 9.7 shows how this was achieved: a swash-plate having its own, built-in angle of inclination was mounted on a Z-shaft machined to a further angle. Rotating one relative to the other varied the net inclination of the plane of the swash-plate to the sliders.

Philips stopped work on Stirling engines in 1978, but licensed a former senior employee, R. J. Meijer, to continue in the United States as STM (Stirling Thermal Motors) with the benefit of *'access to the . . . patents, know-how, and computer simulation programs'* (Hargreaves, *op. cit.*). To that date there had

Fig. 9.6 4-215 DA engine. This was evaluated in a Ford Torino car. Reproduced by permission of Philips Electronics BV

been no disclosure of the thermodynamic design content of that know-how, and Meijer has not, as far as is known, divulged anything since. A costly and unique program of pioneering engineering development has left a legacy of breathtakingly engineered prototypes – but not so much as a hint of the philosophy underlying the hallmark heat transfer and fluid flow design – expansion exchanger with a small number of long, large diameter heater tubes, cooler with a larger number of short, small diameter heater tubes, and regenerator of uniform wire mesh.

STM and United Stirling are continuing to develop engines. These are in the Philips tradition, and are not studied here, since the search for the thermodynamic rationale must focus elsewhere. This chapter therefore closes with a look at two independent developments. Indirectly, these will bring the study closer to its objective – a feel for prospects for the air-charged engine.

9.7 Two UK initiatives

1977 saw the inauguration of an event which was to change once and for all the concept of the 'toy' hot-gas engine. Professor D. H. Chaddock introduced the

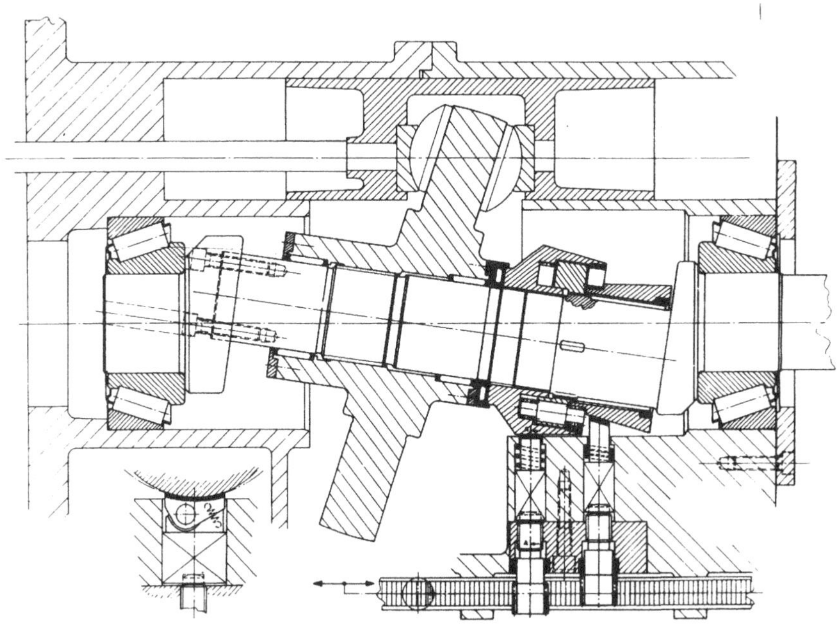

Fig. 9.7 Power modulation for the Advenco engine was by stroke variation achieved by varying the effective angle of the swash-plate. Reproduced by permission of Philips Electronics BV

'Hot-Air Engine Competition' as a feature of the Model Engineering Exhibition, an annual occasion held by tradition at Olympia in London. Model engines with a limit on swept volume, but no restriction on charge pressure or working fluid, were tested on a dynamometer, and the engine of the highest brake output was to be the winner.

With no prior experience of Stirling engines, Eddie Clapham, an engineer at Bristol Aerospace, rose to the challenge with the engine shown at Fig. 9.8. The essence of his approach was to miniaturize the technology of the full-size, Philips-type hot-gas engine. In his own words (Clapham 1999) he had the benefit of 'access to the company scrap bin', and thus to off-cuts of high-duty alloys, stainless-steel locking wire (for the regenerator) and Ampep™, a high-performance dry bearing material used in aircraft control rod ends. The result was very different from the parallel scaling-down exercise of decades previously which had led to toy engines such as that of Fig. 2.17. Charged with helium to 80 bar, a brake output of 95 W was recorded at 2000 r/min from the diminutive 5 cc displacement, and the engine was apparently capable of more.

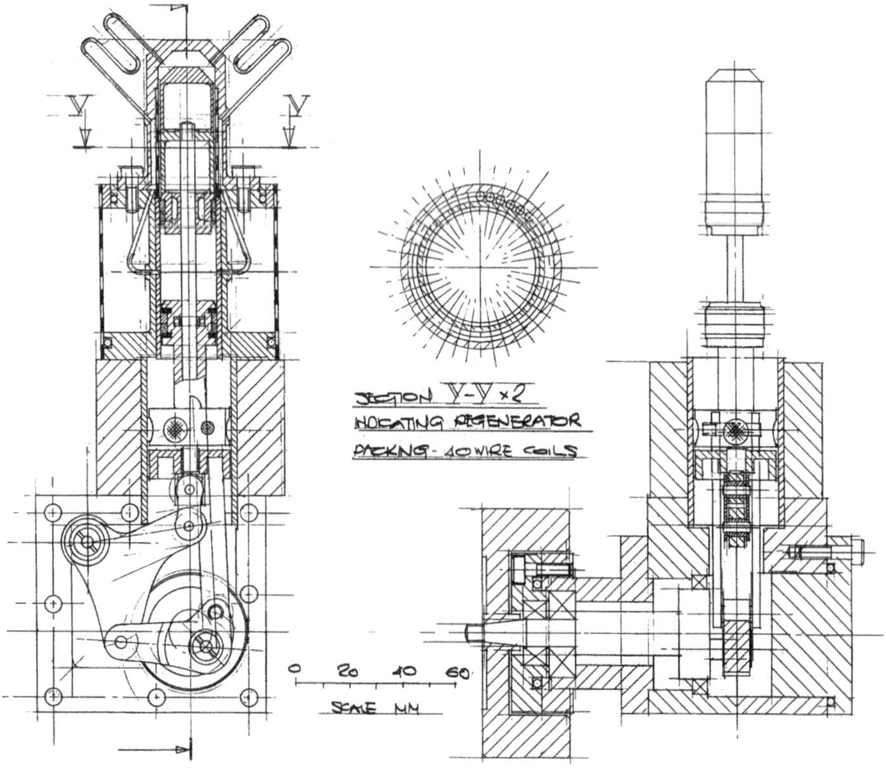

Fig. 9.8 Engine designed by Clapham for operation with high-pressure helium

The engine contained its element of 'technology mismatch', requiring to be brought to high speed by an electric motor before becoming self-sustaining. Calculation of specific cycle work – or Beale number, N_B (Chapter 8) reveals more:

$$N_B = \frac{power}{p_{ref} V_{sw} f} = \frac{78\,[\mathrm{W}]}{81 \times 10^5 [\mathrm{Pa}] \times 2000/60 [\mathrm{s}^{-1}] \times 5.0 \times 10^{-6} [\mathrm{m}^3]}$$

At 0.058 the value is about one-third of the norm (N_B = 0.15–0.2) which might have been achieved had Clapham had access to the principles of gas circuit scaling: the coarse regenerator, spiral-wound from locking wire, afforded about one-tenth of the *NTU* of the typical full-size counterpart. However, the engine had achieved what theoretical deduction alone cannot: it had established the essential point that a miniature Stirling engine need no longer be an impotent toy.

An independent approach to the design of a full-size engine has been taken by Richard Kinnersly of Sustainable Engine Systems Ltd (SES), London. The target application is domestic combined heat and power (CHP) where reliability and low cost are at a premium. Figure 9.9 shows the essential features. The challenge of working fluid containment is approached by designing around nitrogen at a modest charge pressure of 15–20 atm. Transverse loading of the reciprocating seals is eliminated in a patented arrangement which guides each piston on a 'stanchion' tube. The crankshaft runs externally to the pressurized region, and is coupled to the pistons via a rocking lever. The latter penetrates the pressure wall at an oscillating, part-spherical face seal.

Gas circuit design can usually be considered independently of even major mechanical departures. SES has its own, independent approach to the determination of lengths, free-flow areas and hydraulic radii of the flow passages. Overall configuration is 'gamma', so the scaling methods of Chapter 8 – based largely on data from 'alpha' and 'beta' machines – must be applied with circumspection. With this reservation, a tentative, independent exercise in scaling at the gas circuit design study stage supported the SES figures. Innovations in heat exchanger design and manufacture allow the large numbers of compression exchanger ducts called for by operation on air to be realized in practice.

9.8 More on similarity and scaling

A 'thought experiment' may be the best preparation for the material to follow. Figure 9.10 shows a Stirling engine. It is supposed that when the engine is charged to rated pressure with hydrogen, peak power of 10 kW and brake

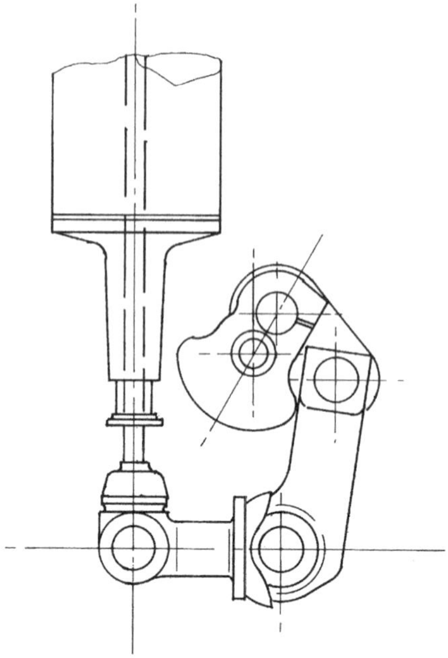

Fig. 9.9 The basis of the SES engine designed for the CHP application, the displacer is guided on a 'stanchion tube'. Piston and displacer actuation are through an arm passing into the pressurized crank-case via a combined seal/spherical bearing. Graphic by Redfern Animation. Reproduced with permission of Sustainable Engine Systems Ltd

thermal efficiency of 30 per cent occur at 2000 r/min. The hydrogen is now released and replaced by nitrogen. With the original exchanger temperatures restored, power and efficiency at 2000 r/min are now (say) 1.5 kW and 19 per cent.

It is suggested that the two cases are best regarded as *two different engines*. Why different? Because at the same charge pressure and r/min power output and efficiency are different. And why are they different? *Because the cycle of gas processes is different*.

A more rational picture of the challenge of Stirling gas circuit design emerges if, for given charge pressure and r/min, the gas processes are required to be (nominally) the same, *regardless of working fluid*. Power and efficiency are then the same, and by the logic now recommended, we have achieved the 'same' engine. The two thermodynamically identical engines have the same swept volume, the same thermodynamic phase angle, α, and the same

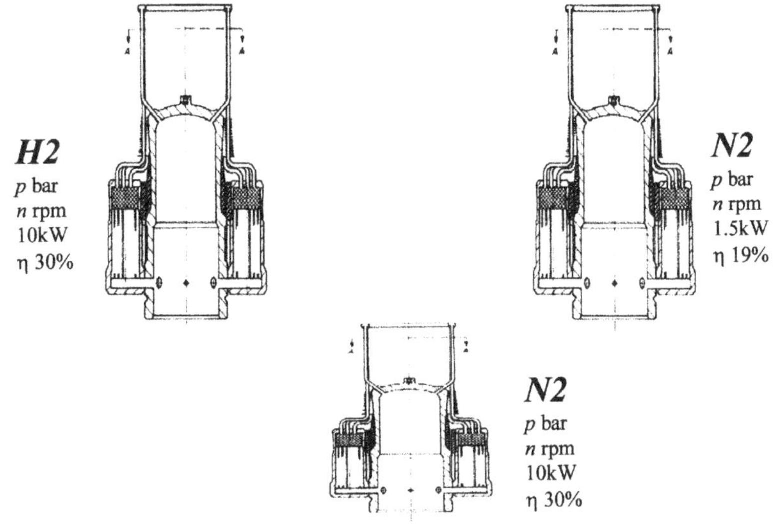

Fig. 9.10 A 'thought experiment' in Similarity principles. Individual images are from an illustration reproduced by permission of Philips Electronics BV

thermodynamic volume ratio, κ. Dead space is distributed as the same fraction of swept volume, but the new engine is otherwise geometrically different (lowermost image of Fig. 9.10) because the heat exchangers have different numbers of tubes. Lengths and diameters are different. The regenerators are of different lengths and free-flow areas. There are differences in wire diameter and mesh of the gauzes.

Achieving similarity of the internal gas processes is part of *thermodynamic scaling*. One of the many reasons why scaling is important is that it allows advantageous thermodynamic features to be carried from one design to another without the necessity for being able to model these features theoretically.

The principles extend without complication where overall size, charge pressure, or r/min (or all three) are required to be different between the two engines. Scaling provides an explanation for one of the most enduring realities of Stirling engine research and development: the engine which produces less than the expected output and/or efficiency. Figure 9.11 shows one of two symmetrical 'modules' of an engine designed for the UK Coal Research Establishment. Each module comprised a pair of gamma-configuration gas circuits, each one of the pair having a displacer cylinder unit connected to opposite sides of a common power piston. Thus there was a total of two double-acting power cylinders and four displacer cylinders. Swept volume per gas

Fig. 9.11 The 26 l, N_2-charged engine designed for the UK Coal Research Establishment (CRE). Only two of the total of four displacer cylinder units are shown installed. Reproduced by permission of the UK CRE

circuit was 5.15 litre, making overall cubic capacity a substantial 20.6 litre and intended for operation at 500 r/min when charged with nitrogen to 56 atm.

On the basis that N_B for the engine should have a value around 0.15, power per gas circuit should be $0.15 \times 56 \times 10^5$ [Pa] $\times 5.15 \times 10^{-3}$ [m^3] $\times$ 500/60 [s^{-1}], or about 36 kW. The concept provides for four gas circuits, or 144 kW total. The illustration shows the symmetrical half of the engine which was tested, and which might therefore have produced 72 kW. A leak from an exchanger joint provides a partial explanation for a measured brake output of only 7 per cent of this value at 5 kW, but the matter can be taken further.

At 5.15 litre, the swept volume of each gas circuit is 5150/61 = 84.4 times the swept volume of the Philips MP1002CA air engine for which the value of N_B reaches (and in fact surpasses) the expected value of 0.15. Charge pressure exceeds that of the MP1002CA in the ratio 56/15 = 3.73. A single gas circuit of the larger engine thus processes 84.4×3.73 or about 315-times the working fluid mass of the smaller engine – although at a cyclic rate reduced in the ratio 500[r/min]/1500[r/min], i.e. a net rate of 105-times greater.

The formal principles of scaling are still to be introduced, but an expectation of 105-times the heat exchange area in each gas circuit of the CRE engine appears warranted. The expansion exchanger of the MP1002CA consists of 160 slots 37.25 mm long, 2.38 mm deep, and 0.4 mm wide. Total wetted area is thus $160 \times 37.25 \times 2 \times (2.38 + 0.4)$ mm^2 = 0.0331 m^2. Each expansion exchanger of the larger engine comprised (Dando 1997) 26 tubes of 442 mm effective length and of 21.1 mm inside diameter. At 0.762 m^2 this is only about one-fifth of the target of 105 x 0.0331 m^2.

The calculation is not to be mistaken for rigorous thermodynamic scaling. On the other hand, there is a tradition of engineering design to take such insight as is on offer, and to use it to bracket the 'true' value when this is unavailable.

10

Thermodynamic design

10.1 The thermodynamic design problem

The underlying difficulty in making sense of Stirling engine thermodynamic design is the large number of *variables* involved. The designer is contending not only with performance (power, r/min, and swept volume) but with operating conditions (temperatures, charge pressure), thermodynamic proportions (phase angle, volume ratio), gas circuit geometry (number, diameter, length of tubes, etc.), and working fluid properties (gas constant, coefficient of dynamic viscosity, thermal conductivity, and specific heat ratio).

The overriding priority must be to bring this number down. Finkelstein's normalization approach has already been introduced (Chapter 8) as a means to a reduction by four variables. It has shown that specific cycle work, N_{TF}, of a Schmidt machine depends *only* on the numerical values of the four parameters N_T, α, κ, and v. That of the more realistic Finkelstein machine is a function of seven parameters – the original four, plus γ, NTU_e, and NTU_c. The trouble is, however, that even this is not (as yet) a *real* engine. It is not possible to carry out laboratory tests with varying NTU_e and plot the result to demonstrate the resulting dependency of thermal efficiency, or of specific cycle work, N_{TF}.

10.2 The task in perspective

It is frequently the case that a novel problem becomes easier to explain when contrasted with a different – but at least familiar problem. In this case, the familiar problem is the basic thermodynamic design of the internal combustion engine. To keep the account manageable, it will suppose the naturally aspirated, four-stroke petrol engine in the unthrottled condition.

The lower part of Fig. 10.1 is a stylized representation of the engine. The upper is the air-standard p–v cycle diagram (ideal Otto), comprising isentropic compression, constant volume heat addition, isentropic expansion and constant volume heat rejection. The cycle diagram scales directly to the ideal indicator diagram of p versus V, since specific volume, v, is linearly related to cylinder volume, V, through the mass, M, of air per cycle.

Corresponding to the four, simple process paths of the diagram there is a simple algebraic statement for ideal indicated efficiency, η_{otto}:

$$\eta_{otto} = 1 - 1/r^{\gamma-1} \qquad (10.1)$$

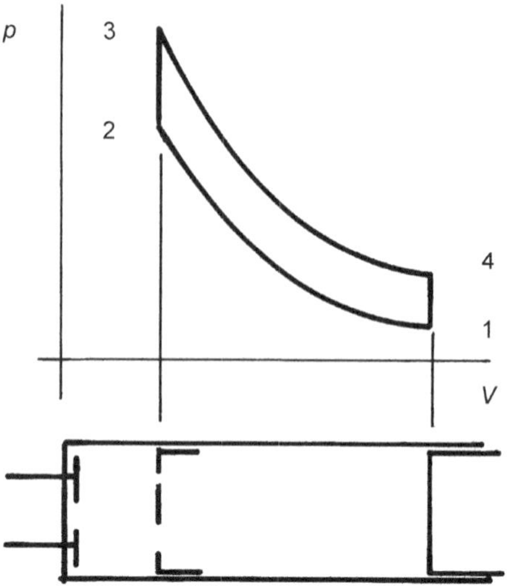

Fig. 10.1 Schematic representation of four-stroke petrol engine and corresponding ideal (Otto) cycle diagram

In equation (10.1), r is compression ratio, v_2/v_1. There is also an expression for ideal cycle work, W_{otto}, but since this evidently takes a different value for each difference size (V_{sw}) of engine, a parameter of more general use is *mean effective pressure*, $\underline{p}_{otto}$, obtained by dividing W_{otto} by V_{sw}. $\underline{p}_{otto}$ has units of pressure, and nothing is lost by a further division by atmospheric pressure to give a expression for specific cycle work precisely paralleling that for the Schmidt cycle. Indeed, if pressure is expressed in terms of bar (or atm) the further division leaves the numerical value unaltered

$$N_{otto} = \frac{r}{r-1}\frac{1}{\gamma-1}\{1-1/r^{\gamma-1}\}\{1/N_T-1\} \tag{10.2}$$

η_{otto} depends only on compression ratio, r, and isentropic index, γ. N_{otto} depends on these, and also on temperature ratio, $N_T = T_3/T_1$. In the jargon, both are *functions* of these parameters. The functional dependence is conventionally written as

$$\eta_{otto} = \eta_{otto}\{r, \gamma\} \tag{10.1a}$$

$$N_{otto} = N_{otto}\{N_T\, r, \gamma\} \tag{10.2a}$$

Equations (10.1a) and (10.2a) mean that, if r and/or γ is changed, then η_{otto} changes. If N_T and/or r and/or γ is changed then N_{otto} changes.

These are gross idealizations: it is common experience that both efficiency and N_{otto} are functions also of air/fuel ratio, *a/f*, ignition advance, ϕ_{ign}, bore/stroke ratio, *b/s*, valve opening and closing angle, ϕ_{ipo}, ϕ_{ipc}, ϕ_{epo}, ϕ_{epc} and, indeed, number of valves per cylinder, n_v (= 2, 3, 4). More realistic forms of the function equations are therefore

$$\eta_{otto} = \eta_{otto}\{r, \gamma, a/f, b/s, \phi_{ign}, \phi_{ipo}, \phi_{ipc}, \phi_{epo}, \phi_{epc}, n_v \ldots\} \qquad (10.1b)$$

$$\frac{power_{otto}}{p_{atm}V_{sw}f} = N_{otto}\left\{N_T, r, \gamma, a/f, b/s, \phi_{ign}, \phi_{ipo}, \phi_{ipc}, \phi_{epo}, \phi_{epc}, n_v \ldots\right\} \qquad (10.2b)$$

There is only one value of γ, namely, 1.4 for air. The rest of the basic design parameters are *simple ratios*, and with the exception of the mass ratio, *a/f*, are *geometric ratios* (an angle is the ratio of two lengths). Additional parameters account for the influence of in-cylinder heat transfer, turbulence, and the combustion processes. These latter effects are largely outside the control of the designer at the outline design stage.

The conclusion, as supported by everyday experience, is that, over a sensible range of sizes, *the internal combustion engine scales geometrically*: an engine of larger or smaller cylinder capacity may be manufactured from a photocopy reduction or enlargement of a working engine (the 'prototype') in reasonable expectation of functioning direct from the drawing board at an efficiency and specific power close to those of the prototype.

At first sight the Stirling engine design problem is different: instead of one value of γ there are two, one each for monatomic and diatomic gases, and as many values of gas constant, R, thermal conductivity, k, and coefficient of dynamic viscosity, μ, as there are choices of working fluid. Whereas there is only one reference pressure for the naturally aspirated internal combustion engine, namely, atmospheric, for the Stirling engine charge pressure may be set at will.

On the other hand, the Stirling engine *does indeed scale*, though not geometrically, in a way which precisely parallels the function equations (10.1b) and (10.2b). Instead of the r and γ of the ideal Otto cycle, there are the four design (or 'scaling') parameters of the (ideal) Schmidt cycle, N_T, κ, α, and the set of dead space ratios, δ:

$$\frac{power_{stirling}}{p_{atm}V_{sw}f} = N_{TF}\{N_T,\ \kappa,\ \alpha, \text{ and the set of } \delta,\ \ldots\} \qquad 10.3$$

The outstanding parameters are going to involve speed, working fluid properties and swept volume. At the time of writing (2000) these have been deduced and their influence on performance conclusively demonstrated by experiment. The complete process of deduction and verification has occupied some years. Presentation here in chronological sequence will amount to a staged exposition of the sought-after picture of thermodynamic design and a corresponding filling-in of the parameter list of equation (10.3).

10.3 Pressure and flowrate

The type of flow under consideration is known as *unsteady, compressible, internal flow*. It is well known that two physical phenomena (at least) play a part in determining the relationship between pressure distribution and flowrate.

(1) viscous effects, sometimes referred to as *viscous dissipation*.

(2) acoustic effects: pressure information travels at local sound speed, a, combined with particle speed, u. A pressure change initiated at a piston face at a given angular location of the crankshaft is not felt at the far end of the gas circuit until the crankshaft has rotated further.

The intensity of viscous dissipation is measured by Reynolds number, N_{re}. In a machine of fixed geometry N_{re} is itself a function of density, ρ, particle speed, $\underline{u}$, and coefficient of dynamic viscosity, μ. For the ideal gas, density is characterized by pressure, p_{ref}, temperature, T_C, and gas constant, R. Particle speed is everywhere determined by angular speed, ω, and crank-throw, r – or indeed *any* length which *characterizes* the machine. In anticipation of extending the treatment to Stirling engines in general, the cube root of swept volume, $V_{sw}^{1/3}$, is chosen for this reference length.

Sonic speed in the ideal gas at temperature T_C, has the value $\sqrt{\gamma R T_C}$. Specific heat ratio, γ is already known to have its own independent effect on the gas processes, so the variables determining the flow phenomena in the motored 'temperature-determined' Stirling engine are T_C, R, ω, μ, V_{sw}, and p_{ref}. There are six variables and four 'dimensions'. By the reasoning introduced in Chapter 8 these are potentially capable of reduction to 6 – 4 = 2 dimensionless groups. A little juggling with possibilities leads to the two dimensionless groupings $\omega V_{sw}^{1/3}/\sqrt{RT_C}$ and $p_{ref}/\omega\mu$. The first is a form of engine speed *independent of working fluid*, and is called *dimensionless speed*, or *characteristic Mach number*, N_{MA}. The second is a fluid-independent charge pressure for which the name Stirling number, N_{SG} has been suggested.

Pressure-related effects are far from the only phenomena active in a Stirling engine operating normally. In order to proceed it is necessary to devise an experiment, or experiments, in which the effects of N_{MA} and N_{SG} dominate, and in which the value of one can be held constant while that of the other is varied.

An engine with both heat exchangers maintained at a common temperature ($N_T = 1$) produces no net thermodynamic work, and, if it is to rotate, has to be driven externally. In this condition it absorbs the power taken to pump the working fluid through the gas circuit. The variation of pressure with volume in the working spaces still generates an indicator (or p–V) diagram in each. In this way there is something to measure corresponding to each combination of charge pressure, r/min and working fluid. To avoid an experimental program of indeterminate duration, it remains to anticipate what to look out for.

It would be possible to instrument almost any Stirling engine and motor it to measure sensitivity to variation in N_{MA} and N_{SG}. On the other hand, if you are focusing on pressure phenomena there is something to be said for exaggerating the conditions in which they occur. All other things being equal, the phase-shift between generated and incident pressure waves, for example, increases with increasing overall length of the gas circuit.

Figure 10.2 shows the SM-1 designed for the purpose. It was essentially of opposed-piston (alpha) configuration. A Ross drive mechanism actuated expansion and compression pistons in identical, parallel cylinders. The neat packaging of the over-length expansion exchanger tubes is the work of Rix (1984), who carried out the work of detail design from the author's specification and arrangement drawing.

Pressure transducers installed at opposite ends of the gas circuit allowed the recording of pressure–time traces, revealing a crucial phenomenon: increasing r/min caused increasing distortion of the trace at the *expansion*-end pressure trace, while the compression-end trace remained relatively unaffected. The relationship between distortion and speed was dependent on working fluid and charge pressure, as might have been anticipated, but there was no obvious reason for the phenomenon *per se*. A computer simulation (Organ 1982) based on the Method of Characteristics (MoC) was available. The MoC is a numerical (and/or graphical) representation of a 'one-dimensional' counterpart of the real wave processes. The computed pressure–time diagrams were found to mimic those from experiment, even to the extent of the additional 'noise' on the expansion trace. However, they did not yield an obvious explanation.

The lack did not matter at the time: the distortion and noise were a sensitive 'marker' for a phenomenon which was obviously dependent on working fluid, speed and pressure. The SM-1 was operated with a variety of working fluids – neon, helium, and nitrogen. It was found that those values of charge pressure and r/min which kept N_{MA} and N_{SG} constant between different working fluids *led to the same distortion and noise of the pressure–time traces*. N_{MA} and N_{SG} were thus shown from both theory and experiment to characterize pumping

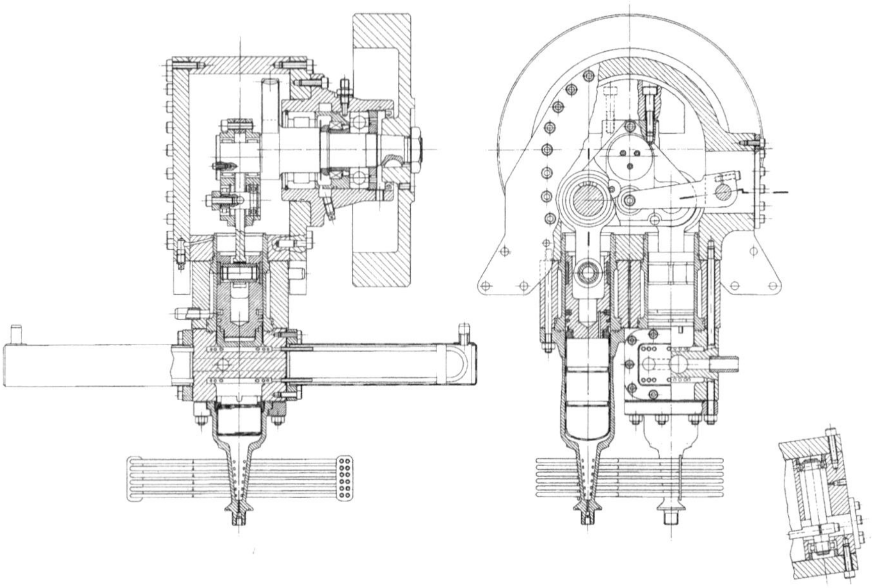

Fig. 10.2 SM-1 machine designed to investigate pressure wave effects. Mechanical design and arrangement drawing by D. H. Rix

losses *and to absorb the effect of different working fluids*. They thus joined N_T as parameters of Stirling engine operation – and therefore of Stirling engine *design*. It was now possible to extend the list of design parameters on which specific cycle work, N_{TF}, depended:

$$N_{TF} = \frac{power}{p_{atm} V_{sw} f} = N_{TF}\{N_T,\ \kappa,\ \alpha,\ \delta,\ \gamma,\ NTU_e, NTU_c, N_{MA}, N_{SG} \ldots\} \tag{10.3a}$$

The list was not by any means complete, but the very fact that N_{MA} and N_{SG} had been identified narrowed the search for the outstanding items.

The objection is sometimes aired that the Stirling engine cycle frequencies and associated piston speeds are too slow for compressibility effects to be important – or, alternatively, that sonic speed, a, is so high that any phase-shift end-to-end is insignificant. After all, $a = \sqrt{\gamma R T_C} = \sqrt{\{1.4[-] \times 287\ [\text{J/kg K}] \times 300\ [\text{K}]\}} = 347$ m/s for air at ambient temperature – and even faster (1315 m/s) for hydrogen with $R = 4120$ [J/kg K]. Do not be misled! At a change in free-flow area an acoustic wave divides into a transmitted component and a reflected component; both components are susceptible to further subdivision. The typical wire gauze regenerator offers a succession of some thousands of area changes –

and the same number of wave reflection sites. The resulting infinity of reflections and re-reflections means that various subcomponents of the original wave arrive late (i.e. with a phase-shift) *while others never arrive at all.*

A problem of such complexity might be thought impossible to quantify. Not at all, provided only that we are prepared to accept a simplification, namely, that $a >> u$ (acoustic speed greatly exceeds particle velocity). *Linear wave analysis* builds on this assumption, dealing with the effect of an unlimited number of reflections and transmissions in a simple algebraic formula applied incrementally to each reflection site. Those interested in the analytical details may wish to consult one of the following accounts: Organ (1993a, b, 1997a), Organ and Rix (1993). The point of importance here, however, is that the parameters of the analysis are precisely the N_{MA}, and N_{SG} already deduced – together, incidentally, with an area ratio to describe the geometry of the gauze.

Figure 10.3 shows a machine, the SM-7, designed to give experimental confirmation of the predictions of the linear wave approach. Oscillations of the piston at up to 7000/min generate pressure waves which echo back and forth through the regenerator stack. The account by Organ and Rix (*op. cit.*) compares experimental pressure–time traces with those generated by linear wave analysis.

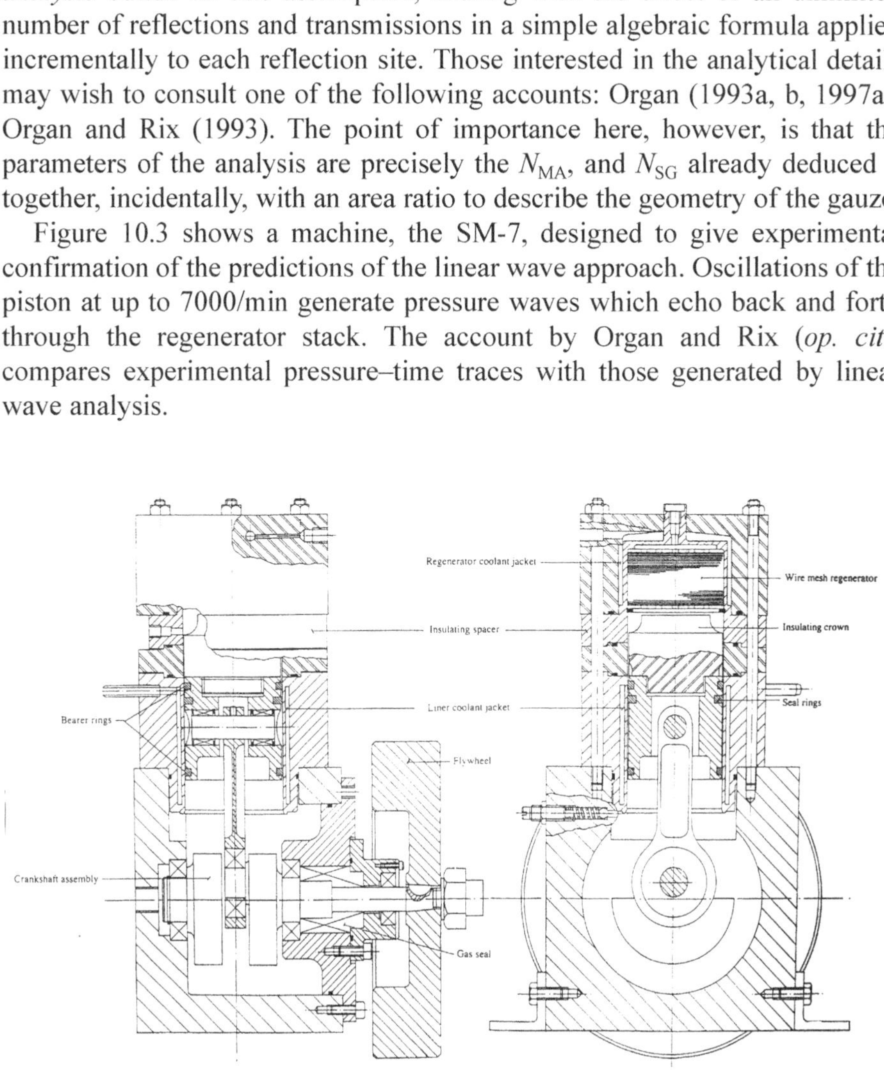

Fig. 10.3 SM-7 machine used to verify the linear-wave analysis of unsteady 'isothermal' flow in the regenerator matrix. Mechanical design and arrangement drawing by D. H. Rix

10.4 Solution of the regenerator problem

Investigation of the pressure phenomena had called for suppression of the usual heat exchange processes. The function of the regenerator, on the other hand, is heat transfer first and foremost, with pressure effects as the incidental, but unavoidable, price. The temperature response had meanwhile held out against both mathematical and computational onslaught, as apparent from a comprehensive review running to an unprecedented 150 pages and 207 references (Kalinin and Dreitser 1994).

When it finally yielded it did so not to computing power, but to an almost trivial change in tactic: all theoretical work since Hausen's initial investigation had been that of the mathematician. In particular, it had dealt separately with the two unknown temperatures – that of the fluid, T_g, and that of the matrix (or 'wall'), T_w. An engineer, on the other hand, sees the phenomenon not first and foremost of *temperature*, but of *heat exchange* – in other words, of *temperature difference* $\Delta T = T_g - T_w$. The engineer is also aware that *heat transfer and friction act on individual fluid particles* (Shapiro 1954), and formulates the expression for heat exchange accordingly, i.e. so that it applies along the particle path (recall Fig. 8.1).

Solution of the resulting equations is as simple as that of the earlier formulations had been difficult, and is accomplished in a single step. The computer plays no part, except to process the solution algorithm and to produce the temperature plots; the basic arithmetic is now within the scope of a hand calculator. Flush ratios of any magnitude are handled with indifference. Figure 10.4 is a set of temperature envelopes from the original solutions (Organ 1994a), one for T_g and one for T_w, and both at cyclic equilibrium. The feature distinguishing the solution for T_g from anything previously seen is the progress of the temperature disturbance which marks the leading edge of the flush

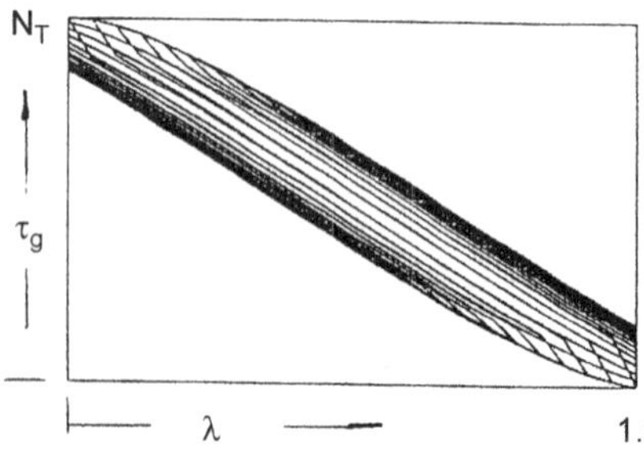

Fig. 10.4 Temperature envelopes from the solutions of the regenerator problem which first accounted for the flush phase

phase. This is perhaps even clearer when displayed in the alternative form of a temperature relief, as in Fig. 10.5.

Temperature solutions are a fine thing – but it appears that two problems remain:

(a) each regenerator design contemplated calls for a separate computation,

(b) the 'regenerator problem' never precisely defined the *in situ* operation in the Stirling engine, and neither, therefore, does the solution.

At face value the first objection holds, but this is not the point: the *parameters* of the temperature solution are those of *all* solutions, and are flush ratio, N_{FL}, thermal capacity ratio, $N_{TCR} = \rho_w c_w T_{ref}/p_{ref}$ and number of transfer units, $NTU_r = N_{str}L_r/r_{hr}$. Specific cycle work N_{TF} [equation (10.3)] therefore now takes the more comprehensive form

$$N_{TF} = N_{TF}\{N_T, \kappa, \alpha, \delta, \gamma, NTU_e, NTU_c, NTU_r, N_{MA}, N_{SG}, N_{FL}, N_{TCR}\} \quad (10.3b)$$

The second reservation is that the NTU_r of equation (10.3b) is a constant whereas, in reality, heat transfer coefficient varies with both time and location. Moreover, the assumptions of the regenerator problem included uniform velocity and density. On the other hand, the new solution *method* takes the extra complications – pressure swing, continuously variable particle velocity, variable heat transfer coefficient, etc. – in its stride. It remains only to embody

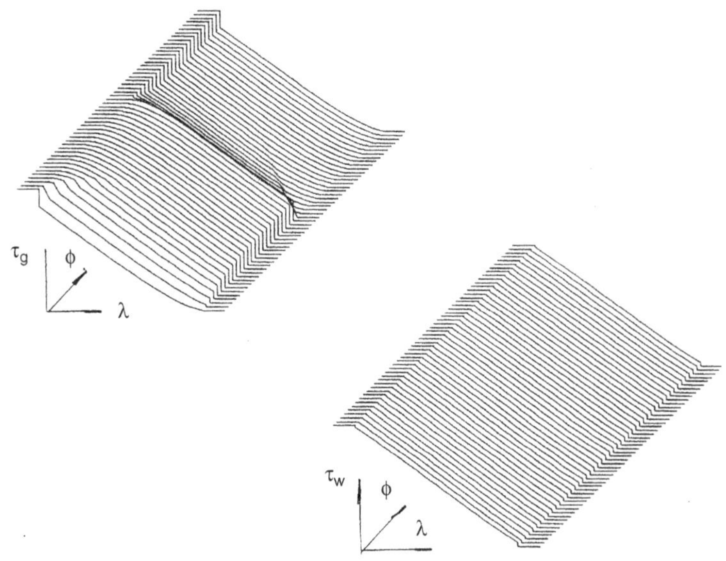

Fig. 10.5 Temperature reliefs for fluid and wall corresponding to the temperature envelope of Fig. 10.4

these variations, or an acceptable approximation to them, in the problem statement, and to reapply the solution method. How this is done will be outlined after accounting for a further important piece of the thermodynamic design jigsaw.

10.5 Gas circuit design by scaling

Emphasis to this point had been on describing the gas processes by algebraic equations and solving numerically by computer. The process is laborious, and has pitfalls: the numerical equivalent of a partial differential equation *cannot*, in general, be properly expressed in terms of fixed increments of time, Δt, and distance, Δx.

However, there are two ways of dealing with equations: to solve, or not to solve! The self-evident advantage of the latter approach is that the analytical description of the gas processes may be made as comprehensive as desired. If this seems too good to be true, it is necessary only to recall (Chapter 8) that *any* equation may be normalized (made dimensionless), and that the process generates two types of symbolic quantity: *parameters* and *dependent variables*, both dimensionless. A simple example is the equation $y = ax + b$: y is the dependent variable, x, the independent variable, and a and b the parameters. The solution (the value of the dependent variable y in this case) depends *only* on the values of the parameters (a and b here). For the Stirling engine, the parameters, and the parameters alone, determine calculated efficiency, specific cycle work, temperature, and pressure histories, etc.

The process offers an instant way of verifying and completing the list of thermodynamic design parameters. Equations for pressure wave effects, convective heat transfer, viscous flow resistance, thermal conduction, etc. are written out in the most ambitious form possible. Further equations relating Stanton number N_{st}, and friction factor, C_f, to Reynolds number, N_{re} for the various flow duct geometries are included for good measure, and all are *normalized* as already described. The process throws up a set of design parameters *which is comprehensive and complete for* an engine defined by the equation set. For the case of an engine in which volume variations are adequately approximated by simple harmonic functions in terms of α, κ, the result is

$$N_{TF} = N_{TF}\{\, N_T, \alpha, \kappa, \quad \begin{array}{llll} \delta_e & & & \\ \delta_{xe} & & N_{Txe} & N_{MA}/\alpha_{ffxe} \\ \delta_r, & \gamma, & N_{Tr}, & N_{MA}/\alpha_{ffr}, \\ \delta_{xc} & & N_{Txc} & N_{MA}/\alpha_{ffxc} \\ \delta_c & & & \end{array} \P_v, N_F, N_{TCR}, N_k \} \qquad (10.3c)$$

Equation (10.3c) is independent of any particular working fluid, charge pressure, or speed. Its use for design is in *scaling*. Scaling may now be clearly seen as the achievement in a new (derivative) design of the essential thermodynamic processes of an engine of known performance (the prototype). The two engines will generally be of different swept volume, different charge pressure, different r/min and different working fluid. Those interested in designing a gas circuit by this method may care to consult one of the comprehensive accounts, e.g. the author's text of 1997, which will explain why some of the parameters of equation (10.3c) have changed form since their earlier appearance. The arithmetic of a gas circuit design is a straightforward task for the hand calculator. A version programmed for PC (*mRT* 1993) is preferable for a succession of exploratory designs.

One way of assessing the scope and limitations of scaling is in terms of an hypothetical exercise in computer simulation: a correctly coded version of the original defining equations (remember, no attempt was made to code them) will return identical numerical values of specific cycle work, N_{TF}, indicated thermal efficiency, and of (dimensionless) pressures, temperatures, and mass flowrates for both prototype and derivative. The two machines will generally be of different power outputs because N_{TF} has to be multiplied up by different values of p_{ref}, V_{sw}, and f (cyclic frequency) in the two cases.

If scaling has already caught your interest, you might like to note the significance of the *order* in which the parameters appear in equation (10.3c). The first three parameters are immediately recognized from the Schmidt analysis. Taken together with N_T, the column of dead space ratios, δ, defines the fourth parameter, v, of the standard Schmidt algebra. We already know that two Schmidt machines sharing common, respective values of these four quantities have the same specific cycle work. Extending the parameter list rightwards amounts to imposing an increasing number of constraints and thus an incremental approach towards detailed correspondence of the cyclic thermodynamic processes of the real machine.

10.6 Similarity principles and engine design

The first engine to be designed throughout using the method is shown in Fig. 10.6. An air-charged design, it was scaled by Larque (1995) from a single gas-circuit of the United Stirling P-40 multi-cylinder, opposed-piston, hydrogen-charged engine using the peak power point as the reference condition. Although of diminutive 30 cc displacement, it embodied full-size engine technology including dry-running piston rings and bearer rings. Apart from the need to rebraze a leaking exchanger tube, the scaled engine ran 'straight off the drawing board' at a no-load speed in excess of 3000 r/min. At the design speed

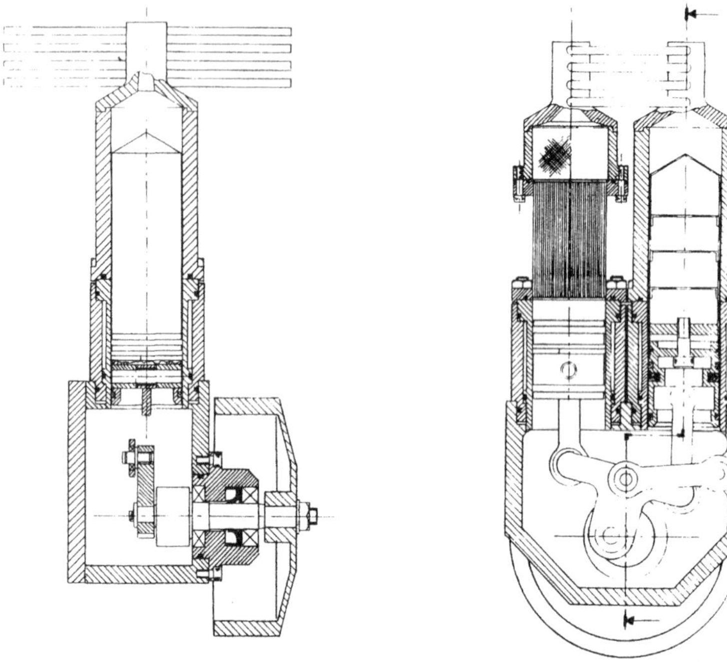

Fig. 10.6 Engine designed by Larque (1995) by scaling from a single gas circuit of the United Stirling P-40 engine. Reproduced by permission of Ian Larque

of 1150 r/min, brake power was 30 W – somewhat down on the scaled value, due possibly to the disproportionate mechanical friction overhead of the single-acting arrangement.

Stirling engine design has been introduced here by contrasting it with that of the four-stroke petrol engine where, over an order-of-magnitude variation in power per cylinder, and despite differences in rated r/min, the various 'gross' design parameters (air/fuel ratio, compression ratio, valve overlap angle, etc.) take nominally the same numerical value from engine to engine. To this extent all four-stroke petrol engines are the 'same'.

The process of 'parameterization' just applied to the Stirling engine has achieved something comparable: all the parameters of equation (10.3c) are dimensionless ratios. What if, in terms of these parameters, all Stirling engines were 'the same' to the extent that internal combustion engines are? Wouldn't that be a shot in the arm for the hard-pressed designer!

Well, they are by no means identical, but between *well-designed*, high-temperature Stirling engines ($N_T \approx 3$) operating at their rated design point, the numerical values of respective dimensionless parameters do not differ greatly. The regenerator heat transfer parameter, N_{Tr}, is defined as $(L_r/r_{hr})^{3/2}N_{SG}^{-1/2}$. If the value is, indeed, consistent from engine to engine, then a plot of $(L_r/r_{hr})^{3/2}$ versus $N_{SG}^{-1/2}$ would be a straight line at 45° on a symmetrical log–log plot; Fig. 10.7 illustrates. The correlation is far from perfect, but spans cylinder powers differing by two orders of magnitude, and to this extent is in line with expectation.

The scaling approach has recently led to a breakthrough in the design of high-efficiency air-charged engines for CHP use. The final chapter is a more appropriate context for a description. In the meantime, it is noteworthy that Japanese investigators (Hamaguchi and Yamashita 1995) have independently begun designing by scaling. Yamamoto *et al.* (1994a and b) found that measured engine performance could be displayed over a wide range when plotted in terms of the similarity parameters.

A second engine proportioned by scaling has both boosted confidence in the method, and also served as a vehicle for characterizing internal pumping losses.

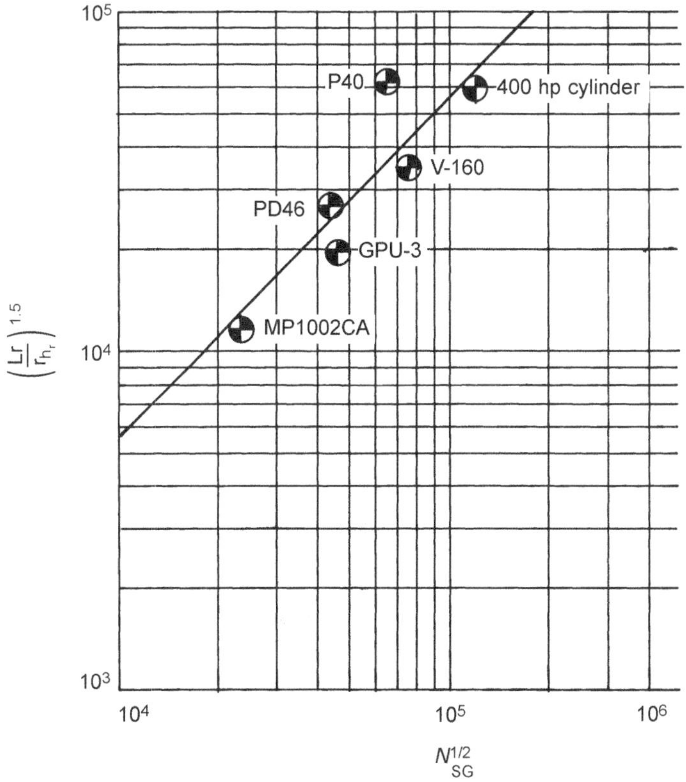

Fig. 10.7 The regenerator heat transfer scaling parameter correlates a wide range of engines. From Organ (1997a) with permission of Professional Engineering Publications

Fig. 10.8 The mRT-25 engine with gas circuit scaled from that of the USS P-40 by Energetic Similarity principles

Figure 10.8 shows the *mRT*-25. Like the Larque engine it was scaled from the USS P-40, but from the peak efficiency point. The different design criterion resulted in a larger number of expansion exchanger tubes of smaller diameter. Rated speed was 1100 r/min with air at 10 atm as working fluid. The Ross crank linkage was used in a pressurized crank-case. Piston rings were of the proprietary PTFE-composite Rulon LD™ and were radially preloaded by O-rings. Bearer rings were of the same material, but not preloaded. With a view to obtaining an accurate energy balance, heating was by passing a heavy electric current through the expansion tubes. A full specification is given in the author's MEP text (1997a).

PTFE-based materials have a temperature slightly above ambient at which the coefficient of thermal expansion increases. This is precisely the range between start-up and operating temperature. There can be a resulting increase in mechanical friction. This and a limitation on electrical input which left expansion temperature some 100° below the rated 700° explain why peak brake power of 30 W fell short of the scaled value of 50 W. Brake thermal efficiency reflected not only these two effects, but also disproportionate thermal conduction down the two, parallel cylinders. Nevertheless, peak efficiency occurred at precisely the scaled r/min value. Organ and Lo (2000) give a fuller account.

Fig. 10.9 Counterbalancing the mass of the dynamometer by suspending it on its axis eliminates both radial and axial forces on the trunnions and allows the accurate zeroing of the torque datum called for in pumping loss tests

The principle of pumping loss measurement involves removing one of the pistons, driving the engine with an electric dynamometer and accurately measuring the mechanical input power over a range of speeds and pressures. With only the piston (or only the displacer) in place the total volume of gas circuit and crank-case (now in open communication) is constant, and the working fluid is pumped between the two at constant total volume, and thus at

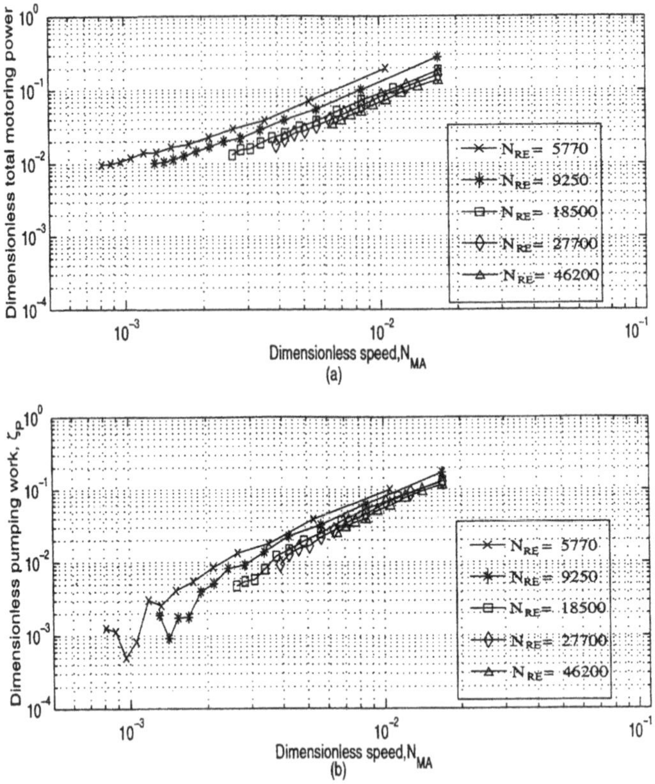

Fig. 10.10 Results of pumping loss tests generalized to any working fluid by plotting against dimensionless speed, N_{MA}, with characteristic Reynolds number, N_{RE} as parameter

nominally constant pressure. The heat exchanger system is then removed and replaced by a dummy gas circuit of similar dead volume, but large free-flow area offering negligible flow resistance, and the tests repeated. In principle, the difference in driving power at a given combination of pressure and r/min is the hydrodynamic pumping power. Correction for the fact that piston friction power is approximately halved calls for separate calibration of main seal friction.

Practical implementation of the principle involved use of a swinging field dynamometer. At the low values of input torque to be resolved, trunnion friction was found to affect the zero reading. The problem was dealt with by mounting dynamometer and engine on a vertical axis (Fig. 10.9) and suspending the dynamometer from a double strand of 25 lb nylon fishing line. The latter passed over a pulley and attached to a weight precisely equal to that of the dynamometer, which was now located by a slight radial clearance instead of the original ball races. With no weight to be reacted vertically, the friction torque was the negligible angular resistance of the nylon line.

Pumping power reduced to a per-cycle value and normalized by $p_{\text{ref}}V_{\text{sw}}$ for direct comparison with specific cycle work, N_{TF}, is denoted ζ_{f}. Like N_{TF} it then becomes a function of the performance parameters N_{MA} and N_{SG}. Theoretical considerations suggest that at constant Reynolds number, ζ_{f} must be proportional to $N_{\text{MA}}{}^2$. Reynolds number in turn is proportional to $N_{\text{SG}}N_{\text{MA}}{}^2$, here abbreviated to N_{RE}. Plotting ζ_{f} against N_{MA} with N_{RE} as parameter in log–log coordinates should result in a set of straight lines. Figure 10.10a shows combined (dimensionless) motoring work, i.e. pumping plus mechanical friction. In Fig. 10.10b mechanical friction has been subtracted leaving ζ_{f}. The results are increasingly ragged at the low-speed end where the subtraction is the small difference between large numbers, but the predicted trend is confirmed. In this format the characteristics are general, and results obtained with air substituted by other gases can be expected to map directly on to the existing curves.

Scaling concepts may now be used to shed light on a question of long-standing: why do the quiet, prepossessing toy hot-air engines return such disappointing performance when 'scaled up' to larger sizes?

10.7 A very un-scale model

For many the toy hot-air engine represents the first acquaintance with the Stirling engine. The toy may have been a proprietary item such as that shown earlier in Fig. 2.17, or it may have been made by a model-maker to one of the many designs described in hobby magazines (Artificer, 1940). In the tradition of such toys, the proprietary item tends to mimic the full-size counterpart – not, however, Stirling's own engines in this case, but the more common domestic or industrial variants of the turn of the nineteenth century. The latter, it has been argued, are not Stirling engines at all, since they dispensed with the regenerator, although the description 'degenerate Stirling engine' might be appropriate.

However, we have just finished arguing that the Stirling engine does not, except for phase angle and volume ratio, scale linearly – either up or down. Being therefore a very un-scale model of a degenerate form, the toy is a long way removed from being a true Stirling engine; but there is worse to come: those who have worked or studied in this area will be familiar with countless proposals to 'scale up' the hot-air engine with a view to achieving a power-producing engine having the now legendary qualities associated with the hot-gas engine – silence, multi-fuel capability, smooth torque, long service intervals, etc. The process is doomed to failure; but there is now an alternative, scientific scaling, either up or down.

The ground is at last prepared for an assessment of prospects for the air engine in the new Millennium. However, it is time to apply the latest theoretical developments to the continuing study of the 1818 engine. Much of the specification has been reconstructed, but it would be reassuring to be able to bracket the estimate of r/min more closely.

10.8 The study of the 1818 engine continued

If the 1818 engine had operated at $N_T = 3$, and been fully developed, equation (8.1) as it stands would have provided a check on the speed estimate. However, it did not and was not. It is still worth calculating N_B for the fact that the value must fall *below* 0.15 on two counts: the crude technology and the reduced temperature ratio. The effect of the latter can be anticipated approximately with the aid of the expression for *ideal*, specific cycle work. This has a salient multiplier $N_T - 1$, consistent with specific work being maximum when N_T is maximum, zero when $N_T = 1$, and negative when $N_T < 1$ (heat pump mode). This suggests dividing the 0.15 of equation (8.1) by the value of $N_T - 1$ using the standard value of N_T, and multiplying by $N_T - 1$ with the value of N_T for the 1818 engine. The revised expectation for N_{TF} is 0.066. Evaluating with f corresponding to 30 r/min of the earlier estimate yields $N_{TF} = 0.2913$, a factor of more than 4 above expectation.

It might be argued that a value on the high side is to be expected on account of the low dead space and the corresponding potential for high compression ratio. On the other hand, when the engine is eventually simulated it will be found that the thermodynamic phase angle given by the complex crank kinematics is unfavourable, reducing compression ratio to the same order as that of modern engines having the dead space associated with tubular heat exchangers.

Either the accepted value of 2 h.p. is a factor of 4 too high, or the speed estimate a factor of 4 too low. As $4 \times 30 = 120$ r/min is three times the estimated flywheel limiting speed, the problem is assumed to lie with the supposed 2 h.p. rating. James Stirling reports (1853) that *'it did not work to the power expected'*, but does not disclose the actual value. An alternative, independent check of the estimated of 30 r/min is evidently desirable.

The appearance of the term N_{MA}/α_{ffr} in the parameter list of equation (10.3b) expresses the fact that regenerator pressure drop as a fraction of charge pressure scales in terms of this ratio. N_{MA} is dimensionless speed, $\omega V_{sw}^{1/3}/\sqrt{\{RT_C\}}$ and α_{ffr} is dimensionless matrix free-flow area, $A_{ffr}/V_{sw}^{2/3}$. Stirling is specific about the dimension of the regenerator annulus (one-fiftieth of diameter) so an assumption for matrix volumetric porosity (say, 0.8) leads to a definite numerical value of α_{ffr}. The Similarity approach says that the numerical value of N_{MA}/α_{ffr} for two fully developed or well-designed Stirling engines operating at respective rated conditions will be numerically similar. With obvious reservation about the state of development of the 1818 engine, the numerical value of the composite parameter N_{MA}/α_{ffr} can be compared with a value of 0.0506 for the Philips MP1002CA at 1800 r/min. The arithmetic gives 2.67 rad/s, or 27 r/min. This is close enough to the estimate via flywheel limiting speed to justify settling on 27 r/min and anticipating a simulated output of close to 1/2 h.p.

The computer simulation output will be more meaningful for an explanation as to how the new regenerator solutions have been extended to the operating environment of the Stirling engine.

11

Completing the picture

11.1 Regenerator analysis further simplified

The new approach has made short work of the 'traditional' regenerator problem. However, in the Stirling engine flow is not uniform between switching and pressure is far from constant. On the other hand, convective heat transfer *still* depends on local, instantaneous temperature *difference*, ΔT, and acts on fluid particles in motion. The extra features are thus readily added to the formulation. All that is required is a representative map of fluid particle motion to replace the idealized picture of Fig. 8.1. How this is done will be outlined later.

Appropriately coded for computer, the comprehensive approach yields (Organ 1997a) the *optimum* regenerator – the combination of volumetric porosity, hydraulic radius, etc. for which the combined losses due to hydrodynamic pumping and imperfect heat exchange at given operating conditions (N_T, N_{MA}, N_{SG}, etc.) are a minimum. A computer-coded search for the optimum is not the same as an *explicit* optimum – a traditional, symbolic algebraic formula expressing ideal flow passage geometry in terms of the parameters of engine operation.

However, it is sometimes the case that, when the appropriate *approach* to a problem has been identified (in this instance, formulation in terms of ΔT), further simplification follows. Used in conjunction with the concept of *NTU* (Number of Transfer Units) and with insights from temperature solutions yielded by the computer-coded implementation, the ΔT formulation allows just this explicit algebraic statement of the optimum. Results can be displayed in the form of charts, bringing design of the optimum regenerator within reach of anyone who can use a hand calculator. To this extent, regenerator analysis has now taken a form of which Robert Stirling himself might have approved. Indeed, it can be convincingly argued (Chapter 12) that the 1818 engine *could not have functioned at all* – let alone pumped water – had Stirling *not* predetermined the wire diameter and winding pitch of the regenerator, possibly along the lines of the material of this chapter. The account is taken largely from the author's paper (2000b) with the permission of the Council of the Institution of Mechanical Engineers.

11.2 Some background

Mean matrix temperature gradient, $\partial T_w/\partial x \approx (T_C - T_E)/L_r$, in the Stirling engine regenerator may be among the most severe encountered in convective heat exchange. With expansion and compression exchangers at 650 and 50 °C respectively, and with a packing length, L_r, of 20 mm, the numerical value is 30 000 K/m. For local temperature difference, $\Delta T = T_g - T_w$, to be a minimum throughout the blow, gas flowing at 10 m/s must change temperature at a rate of 300 000 K/s!

An impression of corresponding thermal loading may be gained from consideration of one of the engines to be studied shortly: the United Stirling P-40 is a four-cylinder, double-acting machine rated (Lundholm 1983) at 45 kW on hydrogen at 150 atm. Thermal efficiency at 4000 r/min and at rated power is 28 per cent. From the ideal gas equation, the mass of working fluid per gas circuit of 95 cm^3 displacement at mean cycle temperature is 150×10^5 [Pa] $\times 95 \times 10^{-6}$ [m^3]/{4120 [J/kg K] × ½(330 + 900)[K]} = 0.00 056 [kg]. A fraction of this mass interacts with the regenerator, doing so twice per cycle. Supposing the fraction to be ½, mass rate is ½ × 2 × 4000/60 [s^{-1}] × 0.00 056 [kg] = 0.0373 [kg/s]. H_2 has a c_p of 14.2 [kJ/kg K], so over a temperature difference of (900 – 330) K, corresponding enthalpy rate at a recovery ratio of 95 per cent is 0.0373 [kg/s] × 14 200 [J/kg K] × 570 [K] × 0.95 = 287 kW! If the value is unexpectedly high, it is consistent with the regenerator's being responsible for some three times the thermal flux of the expansion exchanger. On this basis one-quarter of 45 kW per gas circuit at 28 per cent overall efficiency suggests 1/4 × 45/0.28 ≈ 40 kW thermal input per gas circuit. Three times this rate pro-rated to a half-cycle (heat from regenerator to gas) is 3 × 40/½ = 240 kW – or close to 1 MW for the 45 kW engine.

The thermal efficiency and specific power of the modern Stirling engine reflect temperature recovery ratios, η_T, in excess of 95 per cent. High η_T means that local cyclic excursions of T_g and T_w are small. This in turn implies high *NTU* together with high thermal capacity ratio, N_{TCR}, and frequent switching (small 'flush ratio', N_{FL}). Achieving high temperature recovery is not difficult; doing so with minimum pumping loss is less easy. The Stirling engine calls for a fine balance between the two conflicting phenomena. To date this has been sought through experimentation and intricate computer simulation.

11.3 Flush ratio

A regenerator installed in a Stirling engine interacts with only a fraction, M_i/M, of the total working fluid mass, M, taking part in the cycle. Figure 11.1 indicates this. Of the mass M_i which interacts, only amount M_r, is resident

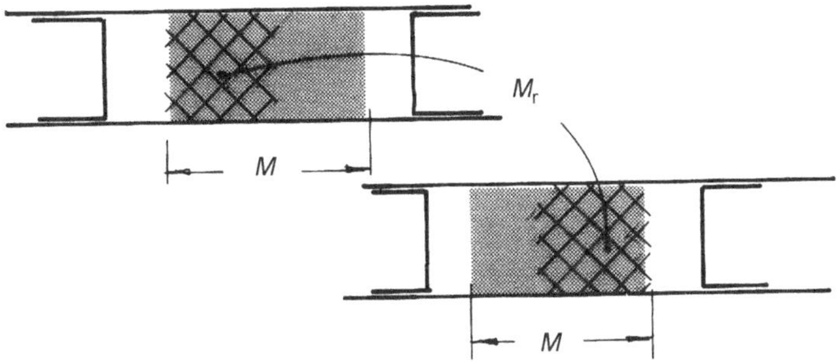

Fig. 11.1 Schematic representation of flush ratio, N_{FL}

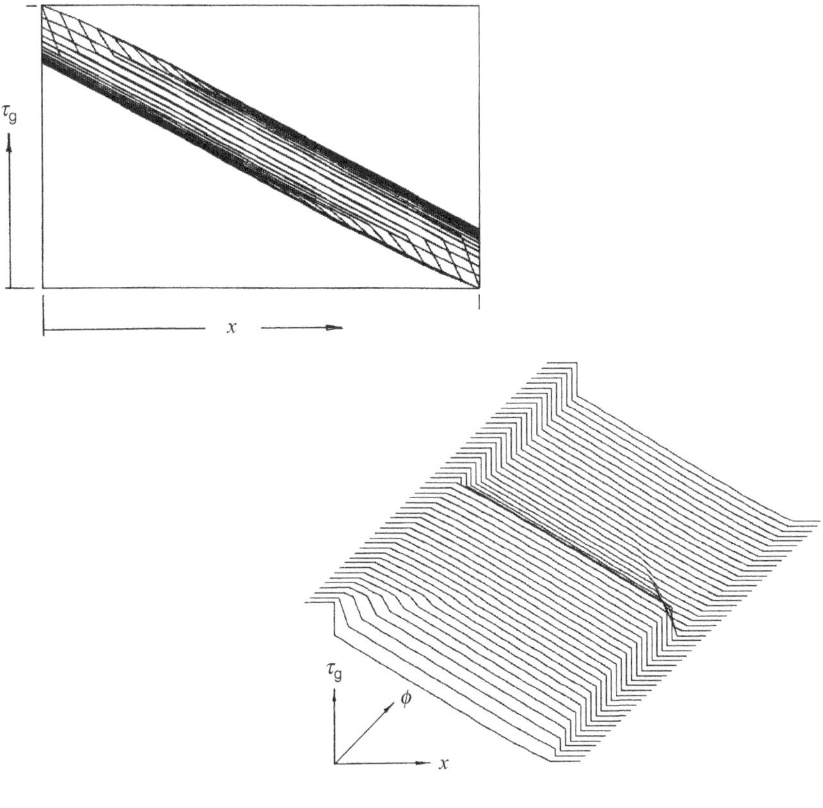

Fig. 11.2 Temperature distribution under conditions of high *NTU*, high thermal capacity ratio, N_{TCR}, and low flush ratio, N_{FL}

within the matrix at any one time, and that amount varies over the cycle due to the pressure fluctuation. Taking an average value of M_r allows definition of *flush ratio*, N_{FL}, as $(M_i - M_r)/M_r$, namely, the ratio of the mass which passes through the matrix to the mass contained within it. Flush ratio, N_{FL}, is an essential link between regenerator theory on the one hand, and Stirling cycle analysis on the other.

A numerical value of N_{FL} for a given Stirling engine may be obtained in terms of the basic Schmidt parameters, the first four groups of equation (10.3c). A table for quick evaluation appears later in this chapter. N_{FL} is not an *additional* parameter, but an *alternative* in the sense that equation (10.3c) may be rewritten with N_{FL} in place of any one of the Similarity parameters N_T, α, κ, or any of the δ used to define it.

The concept of flush ratio is going to feature prominently in the algebraic development. In terms of regenerator theory, where mass rate, m', is uniform, mass throughput $M_i - M_r$ is numerically equal to the product of m' with blow time, t_b, allowing occurrences of the term $m't_b/M_r$ to be abbreviated to N_{FL}.

11.4 Algebraic development

Computer evaluation of fluid and matrix transient response under conditions of high N_{TCR} with small flush ratio has shown (Organ 1997a) that matrix temperature, T_w, remains essentially linear over the cycle, and that the small, wholesale excursions of T_w are as suggested in Fig. 11.2. The intuitive approach to the temperature solutions relies on these conditions being fulfilled. For the moment the remainder of the working assumptions are those of Chapter 8.

11.4.1 Temperature profile

Attention focuses on a slab of fluid dx thick. Figure 11.3a supposes provisionally that heat transfer coefficient, h, is zero. In this extreme case the fluid slab entering from the left with initial temperature T_E proceeds through the matrix with the value unaltered. Temperature *difference*, $\Delta T = T_g - T_w$, at the flow front increases at precisely the rate $-\underline{u}\partial T_w/\partial x$, i.e. at $-\underline{u}(T_C - T_E)/L_r$.

Now suppose that h is finite, so that exchange of heat occurs. The rate of increase in ΔT is now less, because the fluid is adjusting downwards towards the matrix temperature profile (Fig. 11.3b). At the same time, the matrix is gaining heat, and its temperature is adjusting upwards. For constant h the *rate* of this adaptation is proportional, *for both fluid and matrix*, to prevailing temperature difference, ΔT, to h, and to wetted area, *perim*.dx. It is *inversely proportional* to thermal capacity, i.e. to specific heat times density times envelope volume. For the fluid this is ρc_p, times A_{ff}dx.

It is probably easiest to visualize events if wall temperature distribution

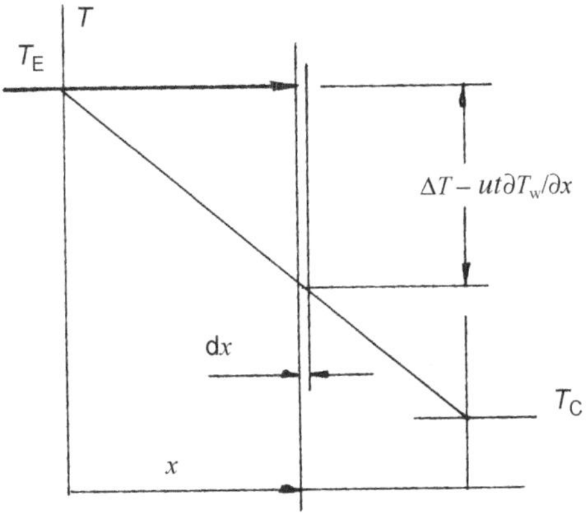

(a) Heat transfer zero

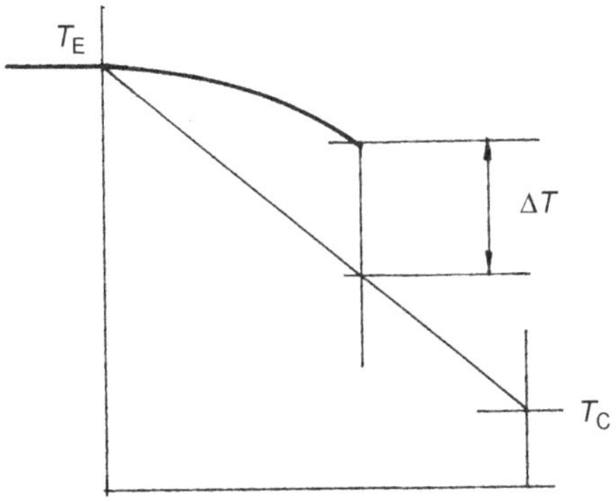

(b) Finite heat transfer

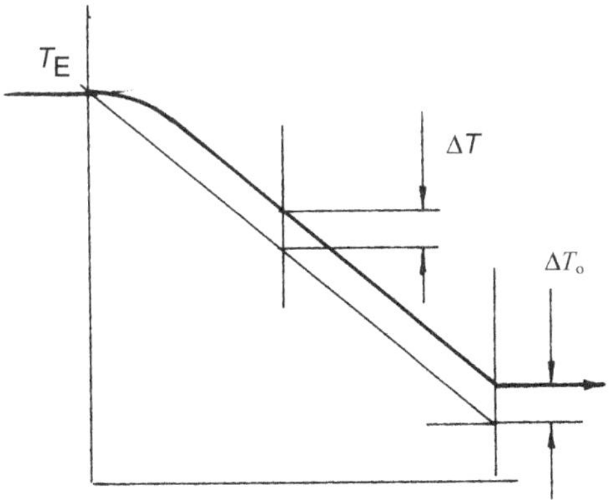

(c) Heat transfer coefficient large

Fig. 11.3 Stages in picture of regenerator operation leading to equations (11.2) and (11.3)

provisionally remains fixed. This applies anyway if thermal capacity ratio approaches infinity, and is the assumption until further notice.

The mathematical operator expressing rate of change when following a moving phenomenon is D/dt, the 'substantial' derivative

$$D\Delta T/dt = -\underline{u}\partial T_w/\partial x - \{h.perim.dx/\rho c_p A_{ff} dx\}\Delta T \qquad (11.1)$$

The ratio $perim./A_{ff}$ is the inverse of hydraulic radius, r_h. But for the velocity modulus term, $|\underline{u}|$, the group in parentheses is Stanton number, N_{st}, divided by r_h:

$$D\Delta T/dt = -\underline{u}\partial T_w/\partial x - \Delta T\,|\underline{u}|N_{st}/r_h \qquad (11.2)$$

Figure 11.3c shows what happens if N_{st} is sufficiently large for useful functioning as a regenerator: after the fluid particle has penetrated by a short distance, the temperature difference, ΔT, due to motion relative to the gradient is sufficient for the resulting heat exchange to balance the tendency for further increase. The particle completes its course through the matrix at rapidly decreasing temperature, T_g, *but at constant local difference*, ΔT. If ΔT is not changing then $D\Delta T/dt$ is zero in turn, and for the remainder of the solution, equation (11.1) becomes

$$\{N_{st}/r_h\}\Delta T \approx -\partial T_w/\partial x \qquad (11.3)$$

Re-expressing $\partial T_w/\partial x$ in terms of the end temperatures, recalling that $N_{st}L_r/r_h$ = NTU and substituting into equation (11.2) gives the final approximation for ΔT. As this is the value at exit from the right-hand end

$$\Delta T_{exit} = T_g^{exit} - T_C \approx -(T_C - T_E)/NTU \qquad NTU >> 1 \qquad (11.4)$$

11.4.2 The 'flush' phase in perspective

The flush phase is initiated by the start of the reverse blow, and may be thought of as combining two, separate transients: first, the temperature-difference profile is translated as far to the left of the matrix gradient line as is required for the heat exchange process to 'catch up' with, and to compensate for, the wholesale convection of the profile. This takes the profile as far to the left of the line of $\partial T_w/\partial x$ = constant as it originally lay to the right. The linear distance, d_{FL}, covered by a particle during this phase is given (Fig. 11.4a) in terms of similar triangles by $\tfrac{1}{2}d_{FL}/\Delta T \approx L_r/(T_E - T_C)$. Substituting equation (11.4) leads to an expression for d_{FL} as a fraction of matrix length, L_r, viz. to $d_{FL}/L_r = 2/NTU$. With blow 'distance' equal to $N_{FL}L_r$ a *readjustment interval* may be expressed as

$$\frac{\text{re-adjustment interval}}{\text{blow duration}} \approx \frac{d_{FL}}{N_{FL}L_r} \approx \frac{2}{N_{FL}NTU} \qquad NTU>>1 \qquad (11.5)$$

According to equation (11.4), when NTU = 40 and flush ratio = 3 the reversal of the profile occupies 1.66 per cent of the blow. Figure 11.4b displays equation (11.5) graphically for quantitative assessment of specific cases.

The second feature of the flush is the temperature front at T_C as the fresh

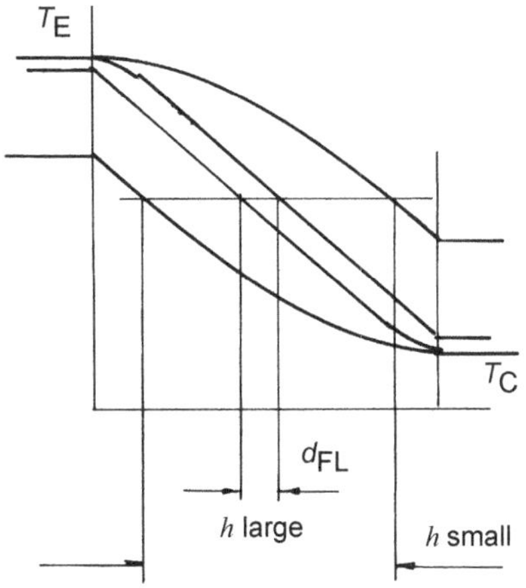

(a) Effective flush ratio is in terms of distance involved in translating fluid temperature profile as far to the left of wall temperature profile as it lay to the right at the instant of switching

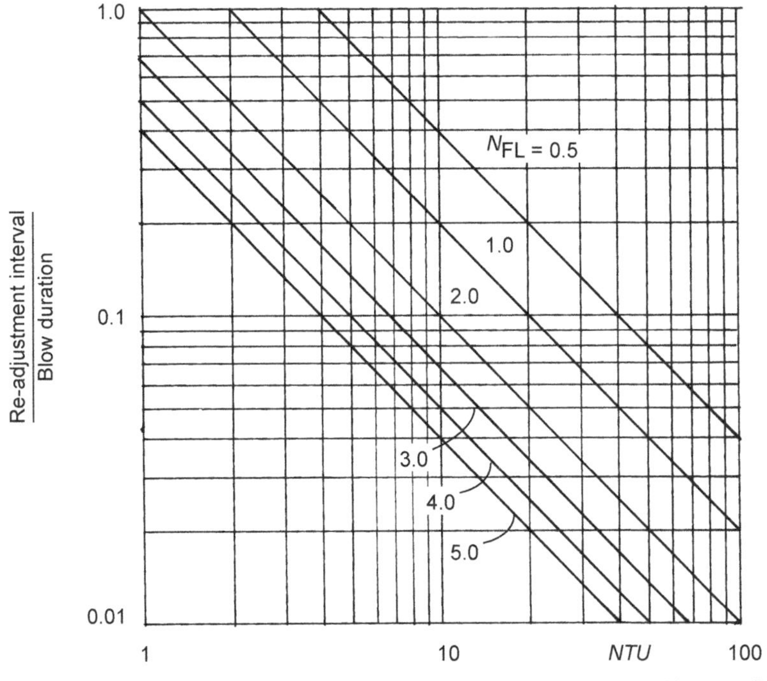

(b) Chart for determination of 'readjustment interval ratio' (effective flush ratio) as function of N_{FL} and L_r

Fig. 11.4 Fluid temperature 'readjustment interval' as a fraction of blow duration (NTU >> 1)

fluid enters. This entry transient takes the full matrix length, L_r, to decay totally, although with high *NTU*, *effective* decay distance is considerably less.

11.4.3 Temperature recovery ratio

A definition of temperature recovery ratio, η_T, is the ratio of temperature change actually achieved at exit to maximum possible temperature change, viz. $(T_g^{exit} - T_g^{in})/(T_C - T_E)$. For the special case of infinite thermal capacity ratio, N_{TCR}, exit temperature swing is that of the fluid alone (i.e. there is no component due to matrix swing). Carrying out the substitution:

$$\eta_T \approx \frac{NTU - 1}{NTU} \approx \frac{N_{st}L/r_h - 1}{N_{st}L/r_h} \quad NTU>>1 \qquad (11.6)$$

For large *NTU* and N_{TCR} the numerical result is indistinguishable from that of Kays and London (1964, their equation 2-13b with 2-24) namely $\eta_T \approx NTU/(NTU + 1)$, obtained via quite different reasoning. Since η_T depends *only* on outlet ΔT, and is independent of inlet transients, there are *no assumptions over and above those of standard regenerator theory* as applied to this special case of infinite N_{TCR}.

According to equation (11.6), η_T may be enhanced without limit by increasing *NTU*. In practice the benefit is increasingly offset by mounting pumping loss. The ideal regenerator is that which strikes the most favourable balance between the two conflicting effects. Matrix temperature swing affects the outcome.

11.4.4 Matrix temperature swing

An energy balance is set up between the fluid element, $\rho A_{ff}dx$ and the matrix element with which it interacts. With $¶_v$ for volumetric porosity, total envelope volume of the combined fluid/matrix element is $A_{ff}dx/¶_v$, of which the solid material occupies volume $A_{ff}dx(1 - ¶_v)/¶_v$. Rate of energy gain by the matrix element is thus $\{\rho_w c_w A_{ff}dx(1 - ¶_v)/¶_v\}\partial T_w/\partial t$, and must be balanced by rate of heat convected, namely, by $h.perim.dx\Delta T$. Equating gives

$$\partial T_w/\partial t = h.perim.\Delta T/\{\rho_w c_w A_{ff}(1 - ¶_v)/¶_v\} \qquad (11.7)$$

The right-hand side may be multiplied and divided by $\rho c_p A_{ff}$ and c_p replaced by $= R\gamma/(\gamma - 1)$. The process throws up the parameter *thermal capacity ratio*, $N_{TCR} = \rho_w c_w T_C/p_{ref}$. The effect on $D\Delta T/dt$ of equation (11.1) is additive:

$$\mathrm{D}\Delta T/\mathrm{d}t = -\underline{u}\partial T_{\mathrm{w}} / \partial x - |\underline{u}| N_{\mathrm{st}} / r_{\mathrm{h}} \left\{ 1 + \frac{\P_{\mathrm{v}}}{1-\P_{\mathrm{v}}} \frac{\gamma}{\gamma - 1} \frac{1}{N_{\mathrm{TCR}}} \right\} \Delta T \qquad (11.2a)$$

Equation (11.2a) is precisely the total differential equation for unsteady fluid temperature resulting from the comprehensive analytical approach (Organ 1997a). Expressing the term in parentheses as $1 + f_{\mathrm{NTCR}}^{-1}\{\P_{\mathrm{v}}, \gamma, N_{\mathrm{TCR}}\}$, treating as per equation (11.2), and bearing in mind that wall temperature at exit is no longer T_{C}

$$\Delta T_{\mathrm{exit}} = T_{\mathrm{g}}^{\mathrm{exit}} - T_{\mathrm{w}}^{\mathrm{exit}} = \frac{-T_{\mathrm{C}}\,(1-N_{\mathrm{T}})}{NTU\,(1 + f_{\mathrm{NTCR}}^{-1})} \qquad NTU >> 1 \qquad (11.4a)$$

Equation (11.4a) is seen to be the original expression for ΔT [equation (11.4)] attenuated by a factor $1 + f_{\mathrm{NTCR}}^{-1}$. The matter of matrix temperature variation will be reconsidered later, when it will be shown, contrary to expectation, *to be independent of NTU!*

11.5 Common denominator for losses

11.5.1 Heat transfer and flow friction correlations

Heat transfer and flow friction effects are correlated for the wire mesh regenerator as for the conventional heat exchanger, i.e. as indicated in Fig. 11.5. This is adapted from the definitive reference (Kays and London *op. cit.*) and shows Colburn j factor (Stanton–Prandtl number, $N_{\mathrm{st}}N_{\mathrm{pr}}^{2/3}$) and friction factor, C_{f} against Reynolds number, N_{re}. For j, an equation which fits closely for any chosen volumetric porosity, $\P_{\mathrm{v}}$, is

$$N_{\mathrm{st}} = a/N_{\mathrm{re}}^{\mathrm{b}} N_{\mathrm{pr}}^{2/3} = a/j \qquad (11.8a)$$

For C_{f}, the underlying physical realities are reflected (Organ 1992a) by a curve of the following form:

$$C_{\mathrm{f}} = c/N_{\mathrm{re}} + d \qquad (11.8b)$$

The curves illustrated are for a specimen volumetric porosity, $\P_{\mathrm{v}}$, of 0.75. Wire screen matrices for Stirling engines have values close to this. Over the range of N_{re} of interest, the curves fit for $a = 0.5$, $b = 0.385$, $c = 80.0$, and $d = 0.35$.

The ratio C_{f}/j versus N_{re} passes through a minimum, and the fact appears to

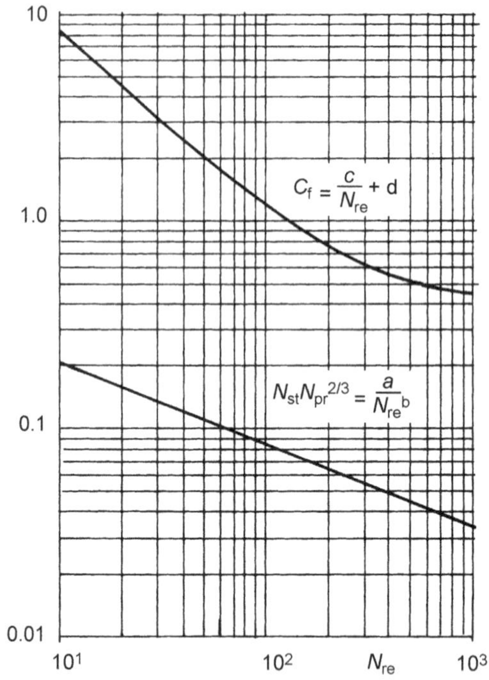

Fig. 11.5 Specimen heat transfer and flow friction correlation fitted to data from Kays and London (1964) for wire screen stack of volumetric porosity, ¶$_v$, = 0.75

be the basis (e.g. Rühlich 2000) for focusing on matrix geometries for which this ratio is an overall minimum; this will not do. The *benefit* of increased $N_{st}L_r/r_h$ and the *penalty* of correspondingly increased C_fL_r/r_h are not linearly related: it takes more than $N_{st}L_r/r_h$ alone to define the penalty of irreversible heat exchange, and more than C_fL_r/r_h alone to define pumping loss. Proper balancing of the two phenomena calls for a common denominator for both: in terms of *lost available work*, T_0dS, heat exchanged across a temperature difference becomes a performance penalty. Seen in this way, a given fractional increase (or decrease) in available work lost to irreversible heat transfer has the same impact on performance as a change in pumping loss of the same magnitude. The optimum regenerator is that whose *NTU* minimizes the sum of the two losses. The emergence of *NTU* as the pivotal parameter allows welcome flexibility in practical design: *NTU* is an abbreviation for the group $N_{st}L_r/r_h$. N_{st} is a function of Reynolds number, N_{re}, which is a function, in turn of r_h. However, L_r is independent of both, suggesting that the required *NTU* may be achieved over a range of operating conditions.

11.5.2 Heat transfer loss

Taking the compression end as an example, excess enthalpy rate at the matrix exit reckoned above nominal exit temperature, T_C, is $H' = m'c_p\{\Delta T + T_w^{exit} - T_C\}$. For the purposes of identifying optimum *NTU* it is not necessary to take explicit account of matrix temperature excess, $T_w^{exit} - T_C$, since it turns out to be independent of *NTU*. Section 11.6 below elaborates. Eliminating c_p in favour of gas constant R and isentropic index, γ, multiplying and dividing by M_r and integrating over a blow of duration t_b gives $H_b = N_{FL}M_r\, RT_o\gamma/(\gamma-1)\Delta T$. ΔT is substituted from equation (11.4), giving

$$H_b \approx \frac{\gamma}{\gamma-1}\frac{N_T-1}{NTU}\frac{1}{(1+f_{NTCR}^{-1})}M_r N_{FL}\, RT_o \tag{11.9}$$

Under present assumptions, H_b has the same numerical value on both hot and cold blows. It therefore represents an extra amount of heat to be added at the expansion end and rejected at the compression end if nominal cycle temperature limits are to be achieved. To this extent it may be thought of as an unproductive 'heat leak' between the cycle temperature limits equivalent to a thermal short. If H_b were processed through the cycle, rather than shorted out, it would ideally yield work output equal to cycle efficiency times H_b, i.e. to $H_b(N_T-1)/N_T$. Seen in these terms, the product of H_b with $(N_T-1)/N_T$ is a loss of potential available work per complete cycle. Carrying out the multiplication, halving to give the loss per blow for compatibility with the rest of the approach and substituting the standard notation, T_oS_q, for available work lost in this way gives

$$T_oS_q \approx \frac{\frac{1}{2}\gamma}{\gamma-1}\frac{(N_T-1)^2}{NTU}\frac{1}{(1+f_{NTCR}^{-1})}M_r N_{FL}\, RT_o \tag{11.10}$$

Apart from the fact that the matrix temperature swing correction, $(1+f_{NTCR}^{-1})$, appears linear rather than squared, equation (11.10) is identical to the result of the standard Availability approach used in the source derivation (Organ 2000c). Under present assumptions (large N_{TCR}) the resulting numerical discrepancy is of little consequence, since the term has a value close to unity for all but small N_{TCR}.

11.6 Hydrodynamic pumping loss

Steady-flow pressure gradient, dp/dx, is defined as $-\frac{1}{2}\underline{u}^2\rho C_f/r_h$. With Q' for volume rate, pumping power per unit length is

$$Q'\mathrm{d}p = |\underline{u}|A_{\mathrm{ff}}\tfrac{1}{2}\underline{u}^2\rho\{C_f/r_h\}\mathrm{d}x$$

Apart from the 'correction' T_o/T_g this is identical to the expression (Organ 1997a) for rate of loss of available work per unit length when mass flows at a rate $\underline{u}\rho A_{\mathrm{ff}}$. Availability theory sees an opportunity for a conversion to useful work where friction reheating takes place at a temperature above the datum T_o. On the other hand, the reheat is scarcely a benefit on the hot-to-cold blow where the objective is to cool the fluid. The simple pumping power expression accordingly appears preferable.

Multiplying and dividing by $\rho^2 A_{\mathrm{ff}}^2$, converting each appearance of $\rho A_{\mathrm{ff}}\mathrm{d}x$ to mass rate m', substituting the surviving ρ^2 by $(p/RT_g)^2$, and approximating T_g by the linear distribution $T_C\{N_T + (1 - N_T)x/L_r\}$ gives

$$Q'\Delta p \approx \tfrac{1}{2}m'^3\frac{C_f L_r}{r_h}\frac{R^2 T_C^{\,2}}{p^2 A_{\mathrm{ff}}^{\,2}}\{N_T + (1-N_T)x/L_r\}^2$$

For the linear temperature distribution assumed, mass content M_r = $\{pV_{dr}/RT_C\}\{\log_e N_T/(N_T - 1)\}$. Substituting and using $A_{\mathrm{ff}}L_r$ for V_{dr} gives

$$Q'\Delta p \approx \tfrac{1}{2}\frac{m'^3}{M_r^{\,3}}\frac{C_f L_r}{r_h}\frac{(\log_e N_T)^2}{(N_T-1)^2}M_r L_r^{\,2}\{N_T + (1-N_T)x/L_r\}^2 \qquad (11.11)$$

With t_b for blow time, uniform m' gives $N_{FL} = m't_b/M_r$. Integrating equation (11.11) with respect to x over the length L_r of the matrix yields a term $(N_T^3 - 1)/3(1 - N_T)$. A further integration with respect to time over the blow for uniform m' amounts to multiplying by t_b. Dimensionless speed, or characteristic Mach number, N_{MA} is substituted for $\omega L_{ref}/\sqrt{RT_C}$, giving

$$Q'\Delta p t_b \approx \tfrac{1}{2}N_{FL}^{\;2}\frac{C_f L_r}{r_h}\frac{(\log_e N_T)^2}{(N_T-1)^2}\frac{(N_T^{\,3}-1)}{3(N_T-1)}\frac{N_{MA}^{\;2}}{\alpha_{\mathrm{ff}}^{\;2}}\frac{\delta_r^2}{\pi^2}M_r N_{FL} RT_C$$

(11.12)

Where equation (11.12) is to be applied to the operating conditions of the Stirling engine, the variation of m' over the blow is arguably closer to simple harmonic than to a uniform rate. For this case it is easy to show that peak m' = $\tfrac{1}{2}\pi\underline{m}'$, where $\underline{m}'$ is the mean for the blow. In this case, $m'(\omega t) = \tfrac{1}{2}\pi\underline{m}' \sin \omega t$, and the integration with respect to time calls for dealing with the term $(\frac{1}{2}\pi \underline{m}')^3\int \sin(\omega t)\, dt$. This is a standard integral with the value $\pi^2/6$ over the interval. Reflecting simple-harmonic velocity variation is a matter of multiplying equation (11.12) by $\pi^2/6$.

11.7 Matrix temperature variation again

The effect of matrix temperature variation has been dealt with rather glibly. It is therefore pursued further to confirm that the effects have been adequately taken into account. The result will be two insights not previously reported.

The constants in equation (11.7) are grouped into the now-familiar parameters. The variation of T_w is supposed linear with time, so integration is straightforward, and leads to

$$\Delta T_w = NTU \Delta T \frac{1}{f_{NTCR}} \frac{\log_e N_T}{N_T - 1} N_{FL} \quad NTU >> 1 \qquad (11.13)$$

Although the analysis does not apply for $NTU = 0$, it appropriately predicts zero variation in T_w over a blow for this case. Substituting for ΔT from equation (11.4a) into equation (11.13) gives

$$\frac{\Delta T_w}{T_C} = \varepsilon_w = \frac{\log_e N_T}{f_{NTCR}\ (1+f_{NTCR}^{-1})} N_{FL} \quad NTU >> 1 \qquad (11.14)$$

Except for being undefined when $NTU = 0$, equation (11.14) is unexpectedly independent of NTU. There are two implications.

(1) Residual enthalpy at outlet, and thus the heat transfer loss component [equation (11.12)], depend on matrix temperature excursion, *but the optimum value of NTU does not!*

(2) For given N_T and N_{FL} (and $NTU > 0$) matrix temperature excursion is a function only of thermal capacity ratio, N_{TCR}, (and of γ and ¶$_v$). This is not an anomaly: high NTU means small ΔT, low NTU means large ΔT. The matrix does not sense the difference.

11.8 Optimum *NTU*

If the matrix were to consist of smooth, parallel channels, it would be possible to invoke Reynolds' analogy and say that friction factor, C_f, when plotted as a function of Reynolds number, N_{re}, parallels the curve of $N_{st}N_{pr}^{2/3}$, i.e. $C_f/j =$

constant (j is Colburn factor). On this basis a constant times NTU may be substituted for $C_f L_r/r_h$ in equation (11.12). With f_{NT} for the term $(\log_e N_T)/(N_T - 1)$:

$$Q'\Delta pt_b \approx \tfrac{1}{2} N_{FL}^{\ 2} f_{NT}^{\ 2} NTU.N_{pr}^{\ 2/3} C_f / j \frac{N_{MA}^{\ 2}}{\alpha_{ff}^{\ 2}} \frac{\delta_r^{\ 2}}{\pi^2} \frac{(N_T^{\ 3} - 1)}{3(N_T - 1)} M_r N_{FL} RT_C \tag{11.15}$$

However, the analogy does not hold for the interrupted flow passage of the typical wire-mesh regenerator. On the other hand, for the purposes of illustrating the algebraic development, the ratio C_f/j for the relatively narrow range of N_{re} typical of Stirling engine operation (~200) does not vary greatly. (For those unhappy with the approximation, the Appendix to this chapter develops an explicit relationship between $C_f L_r/r_h$ and NTU, and shows that the end result is only slightly affected.)

Provisionally replacing $C_f L_r/r_h$ in equation (11.12) by a constant times $NTU.N_{pr}^{2/3}$, pumping loss becomes proportional to NTU, while heat transfer loss [equation (11.11)] is inversely proportional. Figure 11.6 is based on the operating parameters of the USS P-40 engine (Table 11.1) and displays the individual losses in function of NTU. Optimum NTU is that which minimizes the combined loss. For specified operating conditions and regenerator stack length, L_r, it is achieved by varying hydraulic radius, r_h. Of the three lines falling from upper left to lower right, the lowermost is heat transfer loss due to ΔT alone. The next higher curve adds in the loss (independent of NTU) due to matrix temperature excursion. The uppermost curve is the sum of the latter with flow loss (curve increasing linearly from lower left to upper right). The minimum of the sum is the optimum NTU according to the analysis, and is essentially independent of the matrix swing component.

The curve of combined loss has a shallow minimum, suggesting that overall engine performance is relatively insensitive to NTU. The impression is misleading: net indicated cycle work is given by subtracting the combined loss from a notional 'ideal, lossless' value, which may be visualized on Fig. 11.5 as any horizontal line. It is necessary only to imagine an engine with a narrow margin between the 'lossless' cycle work and the curve of combined loss to see that the NTU can be critical. A pertinent example arises later.

The operating NTU for the USS P-40 engine is marked on Fig. 11.6, and will be seen to fall close to the computed optimum. To make optimum NTU explicit, equations (11.11) and (11.15) are added, the resulting sum differentiated and set equal to zero:

$$NTU_{opt} = \frac{6(N_T - 1)^{3/2}}{(1+f_{NTCR}^{-1})} \frac{\sqrt{\{\gamma/(\gamma - 1)\}}}{N_{FL}(N_{MA}/\alpha_{ff})f_{NT}\delta_r} \frac{1}{\sqrt{\{N_T(N_T^3-1)N_{pr}^{2/3}C_f/j\}}} \qquad NTU_{opt} >> 1 \tag{11.16}$$

The relative ease with which an explicit statement of the optimum is acquired affords a possible explanation to a long-standing mystery. This aspect will be returned to later.

Optimization is potentially a treacherous tool – a mindless process which exploits with indifference not only the physical realities, but also the conceptual weaknesses of the analytical model. In any case, there is, as yet, no basis for judging the predictions of equation (11.16), since the *NTU* of Stirling regenerators are not quoted. The safest way of proceeding is on the basis of a feel for the *NTU* arising in service.

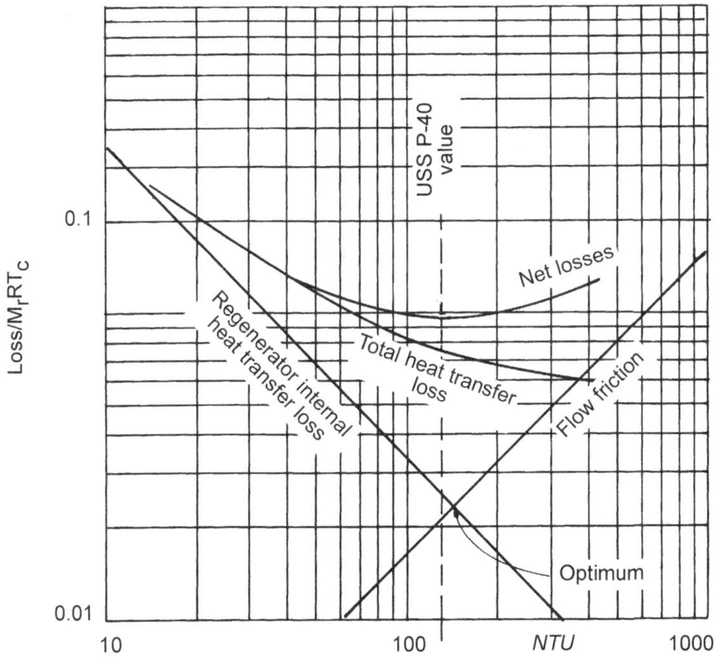

Fig. 11.6 Influence of *NTU* on heat transfer loss, hydrodynamic pumping loss and combined loss for parameters of the United Stirling P-40 (Table 11.1)

Table 11.1 Operating conditions and computed optimum regenerator for six Stirling engines

Parameter	*MP1002CA*	*GPU-3*	*P-40*	*1818*	*400 h.p. /cylinder*	*Clapham 5 cc*
Operating conditions						
$N_T = T_E/T_C$	2.92	3.27	3.07	1.88	3.08	3.07
γ	1.4	1.4	1.4	1.4	1.66	1.66
L_r/r_{hr}	507	754	1636	546*	1540	197
d_w [mm]	0.04 (crimped)	0.04	0.05	1.125*	0.14	0.712
mesh/inch	162 (equiv.)	200	200	8.61*	96.5	—
$\delta_r = V_{dr}/V_{sw}$	0.316	0.461	1.298	0.221	0.298	0.354
$\alpha_{ffr} = A_{ff}/V_{sw}^{2/3}$	0.498	1.0016	1.483	0.0818	1.124	0.237
$N_{SG} = p_{ref}/\omega\mu$	0.564×10^9	2.18×10^9	4.27×10^9	2.08×10^9	14.5×10^9	2.28×10^9
$N_{MA} = \omega V_{sw}^{1/3}/\sqrt{RT_C}$	0.0210	0.0166	0.0191	0.00451	0.0139	0.0046
N_{MA}/α_{ffr}	0.0422	0.0166	0.0129	0.0551	0.0124	0.0194
$N_{TCR} = \rho_w c_w T_{ref}/p_{ref}$	876	193	87.7	10,000	312	166
¶$_v$	0.8	0.75	0.685	0.8*	0.582	0.416
Computed operating conditions						
N_{FL}	3.39	2.72	0.69	4.6	3.58	2.33
$\underline{N}_{re}$ eqn (11.18)	273	184	218	1744	197	549
N_{st} eqn (11.8a)	0.073	0.085	0.079	0.035	0.083	0.0557
NTU	37.04	63.99	130.22	19.51	127.49	10.99
Spurious optimum $\underline{N}_{re}$ (see text)		372 (common to all cases)				
Optima according to present study						
NTU_{opt} eqn (11.16)	37.75	85.63	145.79	16.90	112.29	93.6
$(L_r/r_{hr})_{opt}$ eqn (11.20)	513	929	1772	491	1403	924
opt r_{hr} [mm] eqn (11.20)	0.0545	0.0242	0.0252	1.25	0.0534	0.027
opt d_w [mm] eqn (11.21)	0.0545	0.0322	0.0463	1.251	0.153	0.151
opt mesh/inch	118	250	220	5.17	88	124
$\Delta p/p$ %	1.67	0.041	0.53	2.05	0.47	0.101
ε_w[%] eqn (11.14)	5.7	16.6	6.3	0.04	4.47	2.79
$\underline{N}_{re}$ at optimum NTU eqn. (11.18)	270	149	201	1937	216	117

*Value inferred from computer simulation

11.9 Inference of *NTU* actually achieved

If η_T is known, *NTU* may be estimated from equation (11.6). If a value of Reynolds number, N_{re}, is available, the corresponding *NTU* can be estimated from the $N_{st}N_{pr}^{2/3}$ - N_{re} correlation for the matrix material (cf. Fig. 11.5).

11.9.1 From temperature recovery ratio, η_T

A consistent claim from Philips' early work (e.g. Hargreaves 1991) was achievement of η_T values in excess of 95 per cent. The information can be exploited on inverting equation (11.6) for recovery ratio, η_T, in terms of *NTU*

$$NTU \approx \frac{1}{1-\eta_T} \qquad NTU >> 1 \tag{11.17}$$

NTU of 19 are required to achieve the claimed η_T = 95 per cent. For η_T = 98 per cent, *NTU* = 49: a marginal increase in η_T calls for substantially enhanced *NTU*.

The unexpectedly high values suggest seeking support via values of *NTU* achieved in operating engines. This calls in turn for an estimate of operating N_{re}, from which N_{st} and thus *NTU* follow. As a first step, N_{re} requires to be expressed in terms of the parameters of the present formulation.

11.9.2 NTU from mean cycle $\underline{N}_{re}$

N_{re} is defined as $4\rho|\underline{u}|r_h/\mu$. In the functioning engine, N_{re} varies with both location and crank angle. The apparent discrepancy between actual operating conditions and the assumptions of the present model reduces somewhat on noting that N_{re} may be re-expressed in terms of mass rate as $4|m'|r_h/A_{ff}\mu$. For m' uniform at any instant, the spatial variation is thus seen to be confined to that due to the temperature dependence of the viscosity coefficient, μ.

The link between the regenerator model and reality is flush ratio, N_{FL}, defined in terms of mean mass rate, $\underline{m}'$, as $\underline{m}'t_b/M_r$. Substituting and using t_b = $\frac{1}{2}t_{cyc}$ = 1/(2 × frequency) = π/ω leads to

$$N_{re} \approx 4\frac{\omega}{\pi}\frac{N_{FL}r_h M_r}{A_{ff}\mu}$$

M_r is estimated using the standard expression from the Schmidt analysis, viz. $\{pV_{dr}/RT_C\}\{\log_e N_T/(N_T - 1)\}$. This is substituted and the numerator and denominator multiplied by $\omega^2 L_{ref}^2$ giving

$$\underline{N}_{re} \approx \frac{4}{\pi} \frac{N_{FL}}{(L_r/r_h)} \frac{N_{MA}^2}{\alpha_{ff}^2} f_{NT} N_{SG} \delta_r^2 \tag{11.18}$$

Equation (11.18) makes use of the fact that fractional dead space, $\delta_r = \alpha_{ff} L_r / L_{ref}$. Numerical values of $\underline{N}_{re}$ are based on the value of μ at the mean of T_E and T_C.

The respective N_{FL} values of Table 11.1 have been computed by the algebra of p. 236 of Organ (1997a) which contained errors. Those wishing to calculate N_{FL} for a specific design may obtain a list of errata from the author. Table 11.2 is based on the corrected algebra, and may be used to interpolate N_{FL} from values of dead volume ratios for designs of standard temperature ratio. The *NTU* of Table 11.1 assume rectangular-weave wire gauzes and follow from equation (11.18) via respective L_r/r_h, and then from the correlations of equation (11.8) (or Fig. 11.4). The latter are strictly relevant to the GPU-3 and P-40 machines only, but are the only data available in the open literature.

11.10 Evaluation of optimum *NTU*

Optimum *NTU* [equation (11.16)] is a function of N_T, N_{FL}, N_{MA}/α_{ff}, δ_r γ, $¶_v$, N_{pr}, C_f/j and N_{TCR}. The N_{pr} values are essentially invariant at around 0.75. The N_T of modern engines vary little from 3.0. For $¶_v = 0.75$, the ratio C_f/j is about 10.0. The influence of variations in $¶_v$ is through the term f_{NTCR} and through C_f/j via the heat transfer and flow friction correlations. The influence of f_{NTCR} is slight except at low N_{TCR}, as may be seen from its definition. The effect through the correlations promises to be greater. The results which follow may therefore not apply for $¶_v$ significantly different from 0.75. Otherwise, equation (11.16) is a function of three parameters: γ, N_{TCR}, and the product group $N_{FL}(N_{MA}/\alpha_{ff})\delta_r$. This has an alternative form in terms of the regenerator speed parameter, viz. $N_{FL}N_{Mr} = N_{FL}\omega L_r/\sqrt{RT_C}$. There are essentially only two values of γ, so the full range of NTU_{opt} may be displayed in two charts, one for $\gamma = 1.4$, the other for $\gamma = 1.66$. Figure 11.7a and b is the result; NTU_{opt} is plotted against $N_{FL}\omega L_r/\sqrt{RT_C}$ with N_{TCR} as parameter.

Of the engines in Table 11.1 only the regenerators of the GPU-3 and the P-40 engines comprise woven wire screens, i.e. have heat transfer and friction characteristics described by the coefficients (*a*, *b*, *c*, and *d*) of equations (11.8a) and (11.8b). Results for the MP1002CA, the 1818 engine and the 5 cc design (Clapham 1977) are included as a formality. Entering the charts via the parameter values gives the respective NTU_{opt} listed. These may be compared with the *NTU* values inferred earlier from rated operating conditions. The same

Table 11.2 Table for calculation of flush ratio, N_{FL}, for engines having N_T = 3.0, volume ratio, κ = 1.0 and phase angle α = 90°. δ_r is dimensionless dead volume of regenerator, ($\delta_e + \delta_{xe}$) and ($\delta_{xc} + \delta_c$) are net dimensionless dead volumes respectively at expansion and compression space temperatures.

$\delta_c+\delta_{xc}$: **0.10**

$\delta_e+\delta_{xe}$ \ δ_r:	*0.10*	*0.20*	*0.30*	*0.40*	*0.50*	*0.60*	*0.70*	*0.80*	*0.90*	*1.00*	*1.10*	*1.20*
0.1	5.35	2.82	1.97	1.53	1.26	1.08	0.95	0.85	0.77	0.70	0.65	0.60
0.2	5.69	2.97	2.05	1.58	1.30	1.11	0.97	0.86	0.78	0.71	0.66	0.61
0.3	6.02	3.11	2.13	1.64	1.34	1.14	0.99	0.88	0.79	0.72	0.67	0.62
0.4	6.32	3.24	2.21	1.69	1.37	1.16	1.01	0.90	0.81	0.73	0.67	0.62
0.5	6.61	3.37	2.28	1.74	1.41	1.19	1.03	0.91	0.82	0.74	0.68	0.63
0.6	6.88	3.49	2.36	1.79	1.45	1.22	1.05	0.93	0.83	0.76	0.69	0.64

$\delta_c+\delta_{xc}$: **0.20**

$\delta_e+\delta_{xe}$ \ δ_r:	*0.10*	*0.20*	*0.30*	*0.40*	*0.50*	*0.60*	*0.70*	*0.80*	*0.90*	*1.00*	*1.10*	*1.20*
0.1	6.32	3.30	2.28	1.76	1.44	1.23	1.08	0.96	0.86	0.79	0.73	0.67
0.2	6.60	3.42	2.34	1.80	1.47	1.25	1.09	0.97	0.87	0.80	0.73	0.68
0.3	6.88	3.53	2.41	1.85	1.51	1.28	1.11	0.98	0.89	0.81	0.74	0.69
0.4	7.14	3.65	2.48	1.89	1.54	1.30	1.13	1.00	0.90	0.82	0.75	0.69
0.5	7.38	3.76	2.54	1.94	1.57	1.32	1.15	1.01	0.91	0.83	0.76	0.70
0.6	7.62	3.86	2.61	1.98	1.60	1.35	1.16	1.03	0.92	0.83	0.76	0.71

$\delta_c+\delta_{xc}$: **0.30**

$\delta_e+\delta_{xe}$ \ δ_r:	*0.10*	*0.20*	*0.30*	*0.40*	*0.50*	*0.60*	*0.70*	*0.80*	*0.90*	*1.00*	*1.10*	*1.20*
0.1	7.24	3.75	2.58	1.98	1.62	1.38	1.20	1.07	0.96	0.87	0.80	0.74
0.2	7.48	3.86	2.64	2.02	1.65	1.40	1.21	1.08	0.97	0.88	0.81	0.75
0.3	7.72	3.96	2.69	2.06	1.67	1.42	1.23	1.09	0.98	0.89	0.82	0.75
0.4	7.95	4.06	2.75	2.10	1.70	1.44	1.25	1.10	0.99	0.90	0.82	0.76
0.5	8.17	4.15	2.81	2.14	1.73	1.46	1.26	1.11	1.00	0.91	0.83	0.77
0.6	8.38	4.25	2.87	2.17	1.76	1.48	1.28	1.13	1.01	0.92	0.84	0.77

$\delta_c+\delta_{xc}$: **0.40**

$\delta_e+\delta_{xe}$ \ δ_r:	*0.10*	*0.20*	*0.30*	*0.40*	*0.50*	*0.60*	*0.70*	*0.80*	*0.90*	*1.00*	*1.10*	*1.20*
0.1	8.13	4.19	2.87	2.20	1.80	1.52	1.32	1.17	1.06	0.96	0.88	0.82
0.2	8.35	4.29	2.92	2.24	1.82	1.54	1.34	1.18	1.06	0.97	0.89	0.82
0.3	8.56	4.38	2.98	2.27	1.84	1.56	1.35	1.19	1.07	0.97	0.89	0.82
0.4	8.77	4.47	3.03	2.31	1.87	1.58	1.36	1.21	1.08	0.98	0.90	0.83

0.5	8.97	4.56	3.08	2.34	1.89	1.59	1.38	1.22	1.09	0.99	0.91	0.84
0.6	9.16	4.64	3.13	2.38	1.92	1.61	1.39	1.23	1.10	1.00	0.91	0.84

$\delta_c+\delta_{xc}$: **0.50**

δ_r: / $\delta_e+\delta_{xe}$	*0.10*	*0.20*	*0.30*	*0.40*	*0.50*	*0.60*	*0.70*	*0.80*	*0.90*	*1.00*	*1.10*	*1.20*
0.1	9.00	4.63	3.16	2.42	1.97	1.67	1.45	1.28	1.15	1.05	0.96	0.89
0.2	9.20	4.71	3.21	2.45	1.99	1.68	1.46	1.29	1.16	1.05	0.96	0.89
0.3	9.40	4.80	3.26	2.48	2.01	1.70	1.47	1.30	1.17	1.06	0.97	0.89
0.4	9.59	4.88	3.31	2.51	2.04	1.71	1.48	1.31	1.17	1.07	0.98	0.90
0.5	9.77	4.96	3.35	2.55	2.06	1.73	1.50	1.32	1.18	1.07	0.98	0.90
0.6	9.95	5.04	3.40	2.58	2.08	1.75	1.51	1.33	1.19	1.08	0.99	0.91

concepts and parameter values are used to compute both 'actual' and optimum. This is therefore not a comparison between experimentally determined values on the one hand and the computed counterpart on the other: like is being duly compared with like.

If the rated operating conditions of the engine are maintained, the hydraulic radius required to yield the calculated NTU_{opt} may be expressed in terms of the $N_{st}N_{pr}^{2/3} - N_{re}$ correlation

$$N_{st}N_{pr}^{2/3}\frac{L_r}{r_h} = \frac{a}{N_{re}^{b}}\frac{L_r}{r_h} \tag{11.19}$$

However, N_{re} itself is a function of r_h/L_r [equation (11.18)]. Substituting for N_{re} and extracting r_h/L_r gives

$$\{r_h/L_r\}^{1+b} = \frac{a}{N_{pr}^{2/3}NTU\{(4/\pi)N_{FL}N_{SG}\{\omega L_r/\sqrt{RT_C}\}^2 f_{NT}\}^b} \quad NTU \gg 1 \tag{11.20}$$

Equation (11.20) is general, and so serves to give $\{r_h/L_r\}_{opt}$ in terms of NTU_{opt}. Table 11.1 lists respective optimum r_h for the six engines for comparison with installed values.

For standard values of N_T and N_{pr}, equation (11.20) may be thought of as a function of NTU with the group $N_{FL}N_{SG}\{\omega L_r/\sqrt{RT_C}\}^2$ as parameter. It is plotted in this form in Fig. 11.8 which may be used for evaluating r_h/L_r from the

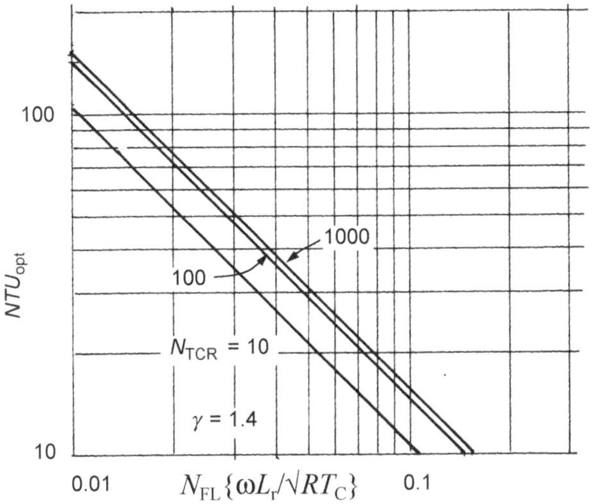

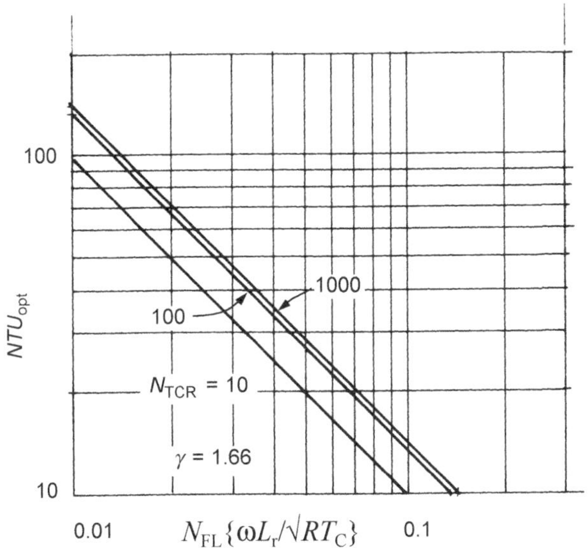

Fig. 11.7 Computed optimum *NTU* as a function of the product group $N_{FL}\{\omega L_r/\sqrt{RT_C}\}$ with thermal capacity ratio, N_{TCR} as parameter. Rectangular screens with N_T = 3.0 and $¶_v$ = 0.75

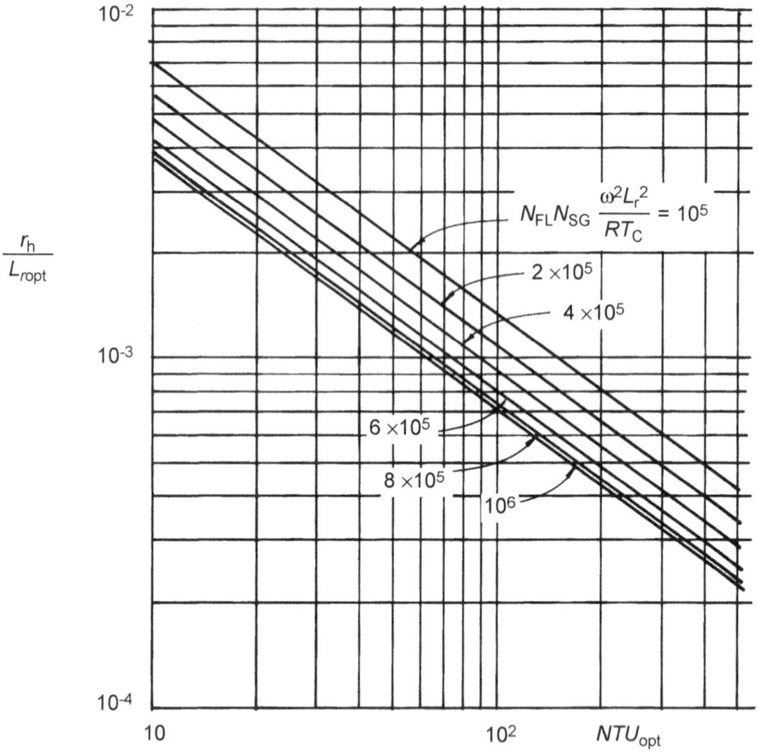

Fig. 11.8 Optimum r_h/L_r as a function of NTU with the product group $N_{FL}N_{SG}\{\omega L_r/\sqrt{RT_C}\}^2$ as parameter. Rectangular-woven wire screens with N_T = 3.0 and $¶_v$ = 0.75.

optimum *NTU* given by Fig. 11.7. It remains to specify wire diameter, d_w, and mesh number, m_w, yielding this optimum r_h.

For stacks of rectangular-woven wire mesh, d_w and m_w relate (Organ 1997a) to volumetric porosity via

$$\frac{r_h}{d_w} \approx \tfrac{1}{4}\frac{¶_v}{1-¶_v} \tag{11.21}$$

$$¶_v \approx 1 - \tfrac{1}{4} m_w d_w \tag{11.22}$$

This completes the design of the optimum regenerator functioning according to present assumptions.

11.11 Implications

It is now possible to investigate the notion (Rühlich *op. cit.*) that the optimum regenerator is that which operates at the minimum ratio of C_f/j. Forming the ratio of the correlation equations (11.8a) and (11.8b), differentiating and setting equal to zero gives

$$N_{re} = c/d\{1/b - 1\} \qquad (11.23)$$

The numerical value for b, c, and d in use here is $N_{re} = 365$. Optima of N_{re} for the modern engines of Table 11.1 vary, but are about half this value. The two approaches are based on common correlations [equation (11.8)]. All N_{re} would therefore have been identical had the two views of the optimum coincided. While there is no doubt much to be said for a matrix geometry which minimizes the ratio C_f/j, the feature in isolation does not make for the optimum regenerator.

The two developed engines of Table 11.1 having woven wire mesh matrices are the GPU-3 and the P-40. Respective optimum *NTU*, according to the present approach, exceed corresponding calculated installed values by 33 and 12 per cent respectively. The small margin in the latter case may reflect the extensive experimental development behind the United Stirling P-40. That its regenerator was close to optimal has already been deduced in an independent approach (Organ 1997a).

Rated and optimum *NTU* coincide almost exactly (± 1.9 per cent) in the case of the MP1002CA. The engine had a filament-wound regenerator, and heat transfer and flow friction correlations therefore somewhat different from those assumed. It remains a pertinent finding that, had the regenerator been of woven wire mesh and of the same r_h and $¶_v$ as actually used, the present treatment would have found it optimum.

'Installed' *NTU* for the fourth engine (Stirling's of 1818) is the value inferred by independent computer simulation (Organ 2000b). This was an exercise in identifying maximum achievable performance, and might have been expected to yield a regenerator closer to the present optimum. On the other hand, the 1818 engine had no heat exchangers. Entry temperature at either end of the matrix was the variable value of the expansion or compression space at any instant. The optimum for those conditions is almost certainly different from that for fixed-temperature entry. Nevertheless, the discrepancy in terms of *NTU* is a modest 11 per cent.

Both the MP1002CA and the engine of 1818 had regenerators of high 'aspect ratio' – a large ratio of flow passage length to free-flow area L_r/A_{ff}. The GPU-3, and particularly the P-40, have the reverse. The expectation is for the former type to be susceptible to higher flow loss. Table 11.1 confirms that the higher the aspect ratio, the lower is the *NTU* value to achieve the optimum balance with pumping loss.

At 0.0463 mm, optimum wire diameter for the P-40 is 9 per cent less than the installed value. The 220-mesh optimum compares with 200 mesh installed. The GPU-3 used 0.04 mm diameter wire at 200 mesh compared with the 0.0322 mm and 250 mesh to achieve optimum *NTU*. For the MP1002CA 'installed' and optimum values coincide.

As far as is known, Philips' 400 h.p./cylinder engine remained a concept, though one which was researched with the aid of the 'part engine'. Hargreaves (1991) gives details of three candidate regenerator matrices. The specification values of Table 11.1 are for regenerator No.2, for which Philips' own estimate of *NTU* (expressed as Hausen's 'reduced length', Λ) was 105. The estimate from the present approach (which puts rated $\underline{N}_{re}$ at 197 versus Philips' 49–102) is 127, with a computed optimum of 112. This is the only one of the four modern engines for which the optimization process indicates that *NTU* should be reduced. The finding is consistent with the uncharacteristically low volumetric porosity (for all three candidate regenerators) for which, however, the a, b, c, and d values of the heat transfer and flow friction correlations would have been somewhat different from those in use here.

The very low *NTU* of the Clapham engine has been deduced independently (Organ 1997a) and attributed to the large diameter of the wire from which the individual coils were wound.

11.12 Complete temperature solutions

The material of this chapter has taken the chronological development of the regenerator temperature analysis out of sequence: in reality, solution of the detailed, transient temperature response under the cyclically varying pressure and flow conditions of the Stirling engine were available in advance of the present, simplified approach. An outline account should be sufficient for an understanding of the comprehensive treatment.

Equation (11.2) is left intact (i.e. the $\mathrm{D}\Delta T/\mathrm{d}t$ term is assumed finite rather than zero) and modified for the effect of pressure swing by the introduction of a term in $\mathrm{d}p/\mathrm{d}t$. It is then solved simultaneously with the equation for matrix temperature [equation (11.7)] at points along the particle paths. The process calls for prior construction of a plot of particle paths, the 'particle trajectory map', to replace the assumption that fluid particle motion is uniform at $\pm\, \underline{u}$ between switching. It is also necessary to superimpose a rectangular grid on the particle diagram whose intersections define fixed points on the (stationary) matrix so that local temperature difference, ΔT, and local matrix temperature,

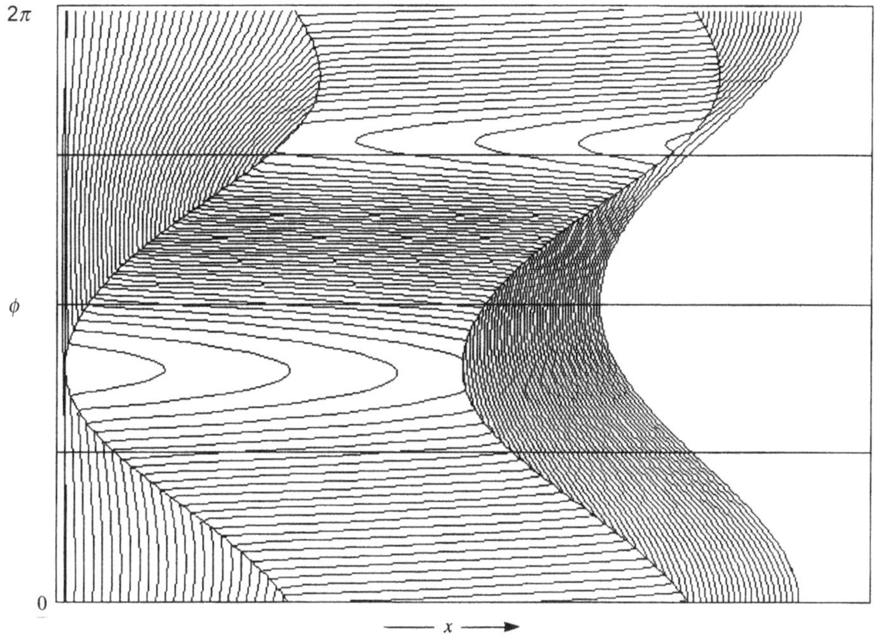

Fig. 11.9 Fluid particle trajectory map for Stirling's 1818 engine with moving regenerator

T_w, may be solved simultaneously.

The work of generating the particle diagrams is more straightforward than Walker's pioneer approach of 1961 thanks to the concept of 'natural coordinates'. This employs *working fluid mass fraction* as a more convenient measure of gas circuit location than the more obvious linear distance: a natural coordinate value of 0.5, for example, is the point in the gas circuit reckoned from the left-hand piston face where precisely one-half of the total working fluid mass, M, is to the left – and the other half, evidently, to the right.

Under the assumption of 'slab' flow (on which all Stirling cycle simulation is based anyway) a working fluid particle, identified by the fact that fraction x of total mass M lies to its left at any instant, *always* retains that same location relative to M throughout a cycle. It may thus be tracked over a complete revolution of the crankshaft. Application to a succession of representative particles (mass fractions, x = 0.1, 0.2, 0.3, etc.) leads to a map of the type shown in Fig. 11.9. This has been constructed for Stirling's engine of 1818, in which the regenerator is in motion. Comparing with Walker's original map for a machine with a stationary regenerator (Fig. 8.3) shows the versatility of the method.

Arranging coincidence of the integration grid calls for selection of

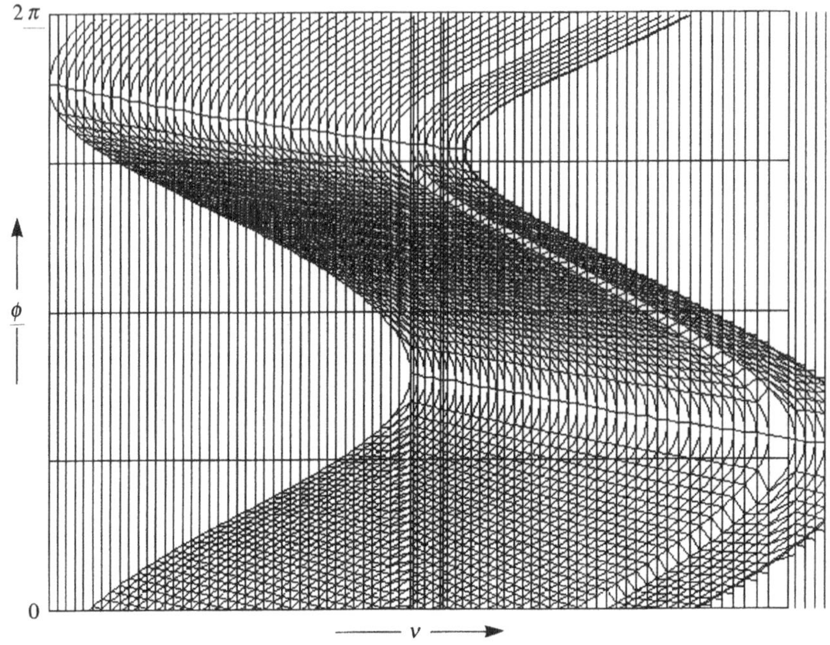

Fig. 11.10 Map for the 1818 engine with superimposed 'fixed' grid forming combined Euler/Lagrange integration mesh

successive mass fractions, x, in such a way that left- and right-hand extremities of the trajectories are tangent to the vertical lines of the grid. The result is the integration stencil of Fig. 11.10, in which the curvilinear Lagrange grid (for the moving fluid particles) superimposes precisely over the rectangular Euler grid (for the 'stationary' matrix) with a certain pictorial attraction.

11.13 Thermodynamic study of the 1818 engine (continued)

The mechanism of the temperature solutions is not confined to the regenerator. It is essentially a statement of the thermodynamic processes throughout the entire gas circuit. To this extent it forms a self-contained cycle simulation. Alternatively, it may be superimposed on an approximate, but readily implemented gas process model, such as Finkelstein's Generalized Thermodynamic Analysis (1960b). Organ (2000a) takes the latter option and the 1818 engine as an example.

The specimen p–V traces of Fig. 11.11 lie about a mean pressure of 1 atm and show the large, clockwise loop for the expansion space, the smaller, anti-clockwise loop for the compression space, and the difference between the two

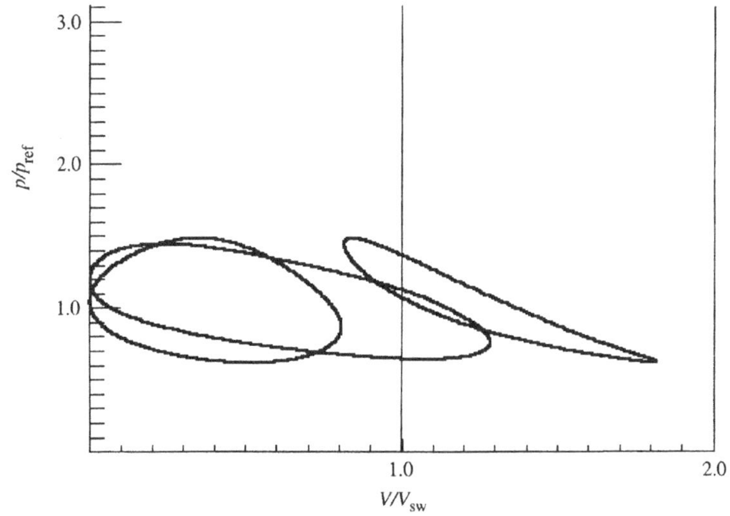

Fig. 11.11 Specimen indicator diagrams for the 1818 engine showing the low pressure swing due to the small temperature ratio and unfavourable thermodynamic phase angle

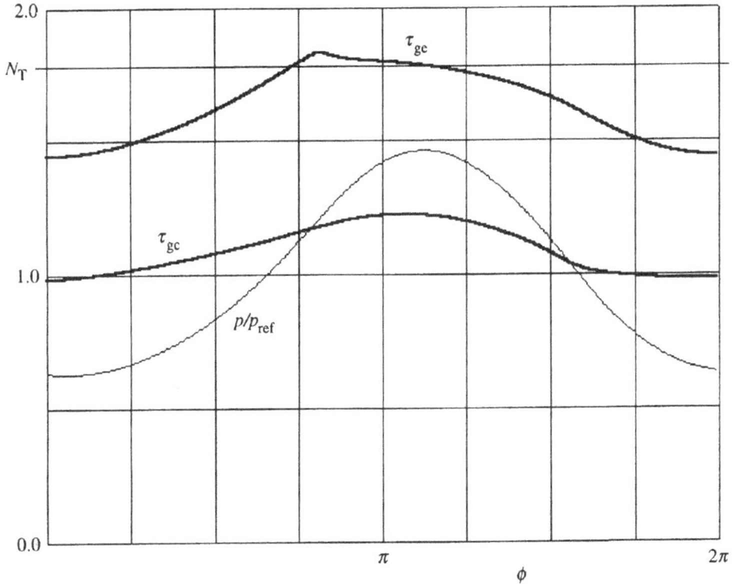

Fig. 11.12 Specimen fluid temperature records for the variable-volume working spaces of the 1818 engine

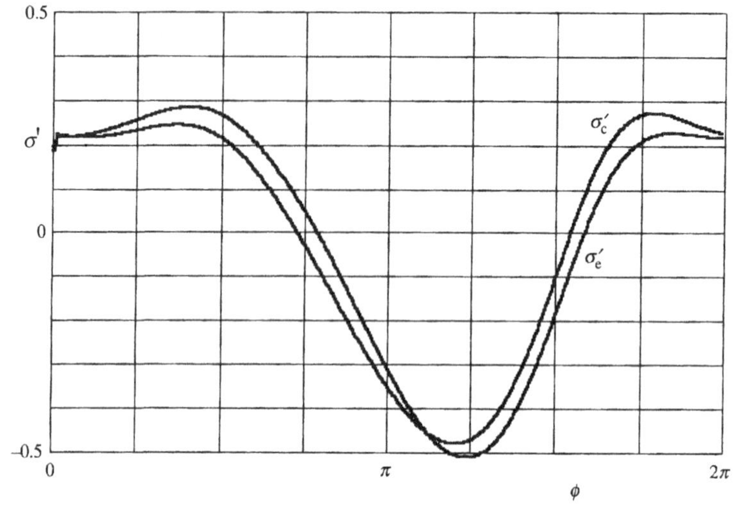

Fig. 11.13 Cyclic variation of working space fractional mass content for the 1818 engine

– the net indicator diagram. The relatively shallow pressure swing might be thought inconsistent with the small dead space. It is due partly to the low temperature ratio, N_T, and partly to the small kinematic phase angle, β, of 38° which corresponds to an unfavourable thermodynamic phase angle, α, of 134°.

Working space temperature swings are appropriately in sympathy with that of pressure, as Fig. 11.12 confirms. In the expansion ($\tau_{ge} = T_{ge}/T_C$) trace the characteristic double peak of Finkelstein's solutions (1960b) is suggested. Temperature variation in the expansion space is biased *below* T_E consistent with a positive flow of heat. The variation in the compression space is biased *above* T_C for analogous reasons.

Figure 11.13 shows dimensionless mass rates at the extremities of the regenerator. The traces will be seen to have much in common with corresponding plots achieved in Creswick's independent look (1965) at the 'isothermal' cycle model where, however, N_T took almost twice the present value. The reduced phase shift between expansion and compression traces is consistent with the small regenerator dead space of the 1818 engine (small regenerator 'capacitance').

The same, basic principle already developed for the optimum regenerator is applied to computation of heat transfer loss and pumping loss in the comprehensive model. However, the two conflicting losses are evaluated locally at each increment of crank angle rotation and summed over a cycle. The process thus reflects the variation of Stanton number and friction factor with time and location.

11.14 Interim deductions

Availability of the simpler, intuitive temperature solutions means that the designer is no longer dependent on computer-coded models for design of a regenerator of near-optimal specification. On the other hand, it should be clear what is meant by such an 'optimum'. First and foremost it is a 'local' value in the sense that dead volume, V_{dr}, free-flow area, A_{ff}, and flow passage length, L_r, (and, indeed, operating conditions), *are prescribed*, possibly as a result of 'scaling' the gas circuit design from an engine of known performance.

Different starting values of V_{dr}, A_{ff}, and flow L_r will almost certainly lead to a different result – to a different 'local optimum'. The search for the 'global optimum', the best gas circuit equipped with the best regenerator, remains for the time being a task for the computer.

Appendix to Chapter 11

Relationship between N_{st} and C_f deduced from experimental correlations

The definitive heat transfer and flow friction correlations of Kays and London (*op. cit.*) are fitted for $¶_v = 0.766$ by equations (11.8a) and (11.8b) respectively of the main text. From equation (11.8a)

$$N_{re} = \{a/N_{st}{}^{b}N_{pr}\}^{1/b}$$

Substituting into equation (11.8b) gives

$$C_f = \frac{c\left\{N_{st}N_{pr}{}^{2/3}\right\}^{1/b}}{a^{1/b}} + d$$

The substitution required by the analysis in the text is for C_fL_r/r_h in favour of $N_{st}L_r/r_h$. Multiplying through, using $NTU = N_{st}L_r/r_h$, and rationalizing gives

$$C_fL_r/r_h = L_r/r_h\{[(c/a^{1/b})NTU^{1/b}N_{pr}{}^{2/3b}](L_r/r_h)^{-1/b} + d\} \qquad \text{(A1)}$$

Equation A1 is a function of *NTU* and may be substituted for the simple, linear version used in the text. Differentiation to define the optimum is straightforward, resulting in a modified equation (11.16). Numerical results are little changed.

12

By intuition, or by design?

12.1 An anomaly

Compared with the treatment of other contributors to the scientific heritage, the literary coverage of Robert Stirling, his regenerator, and engine, is unbalanced. Accounts by others of their own intellectual struggle with the regenerator problem is dauntingly large; there is comprehensive coverage of the various programs of twentieth-century engine development, and over the past two decades, Stirling's life and ministry have attracted increasing interest. However, whereas it is the norm for scientific contributions in other areas to be scoured for the slightest hint as to how the imaginative step was taken and implemented, Robert Stirling has not, thus far, attracted his fair measure of attention.

Here was no ordinary man: having studied advanced Latin and Greek, Logic, Metaphysics, Mathematics, Rhetoric, Hebrew, and Physics at university level, becoming ordained and running a parish ministry, he turned his hand to technology and to an important invention. And not just *any* important invention, but one which

(a) was to have a measurable impact on the industrialization of Europe through massive economies in metal smelting;
(b) was to make cryogenic engineering possible;
(c) has been judged (Jenkin 1885) '*one of the greatest triumphs of engineering invention*';
(d) was to make possible the regenerative gas turbine;
(e) remains a concept so subtle that it has taken 180 years to give a full scientific analysis of the temperature response.

If his was a sideline commitment to engineering it was without precedent for thoroughness. Current UK patent law calls for the specification to be illustrated by an 'embodiment', a specific instance of the application of the inventive step. It is not clear whether this was a requirement in 1816, but the embodiment he proffered was nothing short of a further invention in its own right – a completely novel prime mover. Yet, again, not just any prime mover, but one

(a) which was on the same grand scale as the work of the full-time engineers of the era – Watt, Newcomen, and Stephenson;
(b) which, when shown to trained engineers dismantled or in the form of an

engineering drawing, is usually not identified as an engine at all!
(c) whose principle of operation is so close to thermodynamic perfection that, when driven backwards, it functions as refrigerator or heat pump;
(d) which was sufficiently sound in concept and realization to go straight off the drawing board into service pumping water.

A further, current requirement of an invention submitted for UK patent cover is that it should be '... *capable of being made or used in some kind of industry*'. Whether driven by legal requirements or his own standards, Stirling complied by pursuing the design of his 'preferred embodiment' to the last bolt and rivet. Even by this point, his engineering design standards are exceptional, as may be seen by contrasting the Scottish specification and drawings with those of the Toshiba patent discussed in the next chapter. Stirling, of course, did not stop at patent and design: as if to dispel any lingering doubt about the new concept, the engine was constructed and put to work draining an Ayreshire quarry.

By any standards this is a success story. In another context the pundits would be speculating on the mental processes leading to the inventive step(s). The 10-foot height of the engine would be scrutinized for significance. Does it reflect a target power output which has been carefully calculated? Does it connote 'scaling' relative to the cubic displacement of other prime movers of the era? Was this a pilot project for an even larger concept? Above all, there would be a quest to establish whether the instant functioning reflects intuitive genius, luck, or routine engineering calculation.

The overdue enquiry will be initiated here. It will come to a view which has not previously seen the light of day, and may change the accepted perception of Robert Stirling the engineer and of his approach to his work.

12.2 The 1818 engine and the regenerator

With the exception of a small problem, the state-of-the-art computer simulation which has already produced the indicator diagrams and particle trajectory maps may be used to examine the regenerator design of the 1818 engine and perhaps read Stirling's intentions. The problem is that the wire diameter and the pitch of the winding are not known. However, by assuming that the heat transfer and flow friction characteristics of the lay-up are not too different from those of woven wire screens, the simulation may be used to find what it would have taken by way of a regenerator for the engine to have functioned at all! It is this process which is going to insist on a radically new interpretation of Stirling as the engineering designer.

Stirling's description is of a lay-up as in the inset of Fig. 12.1a, with the straight wires aligned with the displacer axis separating those wound circumferentially. In order to proceed it is necessary to work with the concept of *hydraulic radius*, r_h, already defined as *wetted volume* of the matrix divided by *wetted area*. For wire screens and filament-wound matrices, r_h may be thought of as proportional to pore size. Stirling was evidently aware of an important relationship between r_h and matrix length, L_r. Figure 12.1b illustrates this: the left-hand diagram shows temperature and pressure gradients across a matrix of length L_r and pore size r_h. In the right-hand diagram, both L_r and r_h are halved, keeping the ratio r_h/L_r constant. At the Reynolds number at which the 1818 regenerator operated (about 2000, see Table 11.1) temperature and pressure gradients are little affected by the change.

The dimensionless ratio r_h/L_r is an important parameter of regenerator design. It is essentially Stirling's own performance parameter: '*... when the width of the passage (r_h) cannot be sufficiently diminished, I increase its length (L_r) to achieve the same end*' (Stirling 1816).

The exercise in regenerator optimization (Chapter 11) was in terms of r_h/L_r. In Fig. 12.2 the horizontal axis is r_h/L_{ref}, i.e. the ratio of r_h to reference length, L_{ref}. For an engine of fixed displacer length, the ratio r_h/L_{ref} is directly proportional to Stirling's variable r_h/L_r.

The curves show the result of varying hydraulic radius at two values of crankshaft r/min, 28 and 40 r/min. The curve ascending from left to right shows heat transfer loss per cycle increasing with increasing r_h/L_{ref}. That descending from left to right shows pumping loss decreasing. Packing length, L_r, is, of course, constant for this particular engine, so the third curve shows net work per cycle as it varies with Stirling's parameter r_h/L_r.

Peak computed indicated power of 588 W (0.79 h.p.) is achieved at 28 r/min, making an interesting comparison with the estimate of 27 r/min via the earlier Similarity arguments on which the provisional specification was based. The value of r_h/L_{ref} at this peak power is 0.0024. Brake output depends upon the supposed mechanical efficiency: at 75 per cent it would have been 0.6 h.p., corresponding to a water pumping capacity at 6 ft head of 18 000 gal/h.

Computed power output falls off precipitously as r/min increases or decreases either side of the optimum: at the 45 r/min estimate of flywheel-limited speed, it has fallen close to zero. Free-flow area as a fraction of reference area ($\alpha_{ff} = A_{ff}/V_{sw}^{2/3}$) is uncharacteristically small, the ratio of flow path length to reference length ($= L_r/V_{sw}^{1/3}$) uncharacteristically large. With cyclic pumping loss proportional to $\{N_{MA}/\alpha_{ff}\}^2$, the rapid fall-off of indicated output reflects 'choking' as speed increases above the rated value. All this suggests a narrow 'window' of both r/min and regenerator design outside which the engine would have been unsuitable for its water pumping duties. An L_{ref} of 0.469 m gives an optimum r_h of 0.0024×469 mm = 1.125 mm or 0.0443 (about 3/64) in. How this value might have been achieved in terms of wire

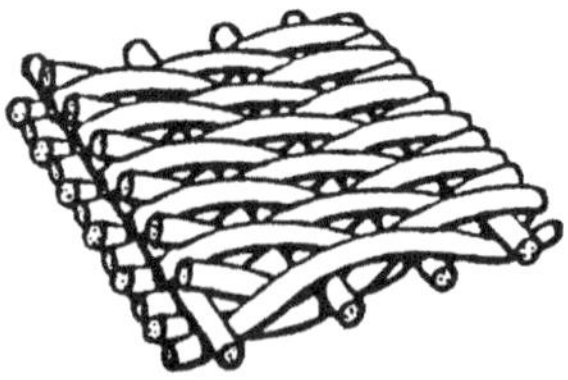

(a) Stirling's description of his regenerator lay-up corresponds to twilled Dutch weave

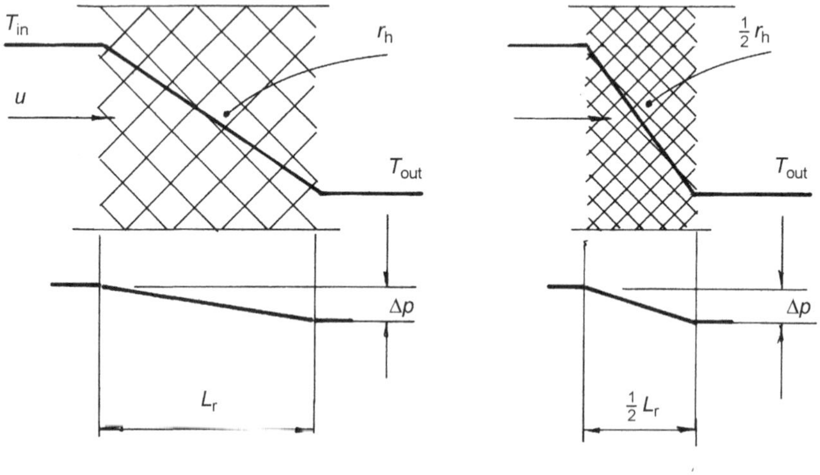

(b) Notation for illustration of Stirling's dimensionless geometric design parameter, r_h/L_r

Fig. 12.1 Stirling's regenerator lay-up, and a schematic representation of the near-constancy of pressure and temperature differences with invariant r_h/L_r

diameter and lay-up would be of interest.

With d_{warp} and d_{weft} for the wire diameters, and m_{warp} and m_{weft} for the reciprocal of respective pitches, volumetric porosity, $\P_v$, and hydraulic radius, r_h, of close-wrapped layers of Dutch twilled weave are approximated by

$$\P_v = 1 - 1/4\pi \frac{m_{warp} d_{warp}^2 + 2m_{weft} d_{weft}^2}{d_{warp} + 2d_{weft}} \tag{12.1}$$

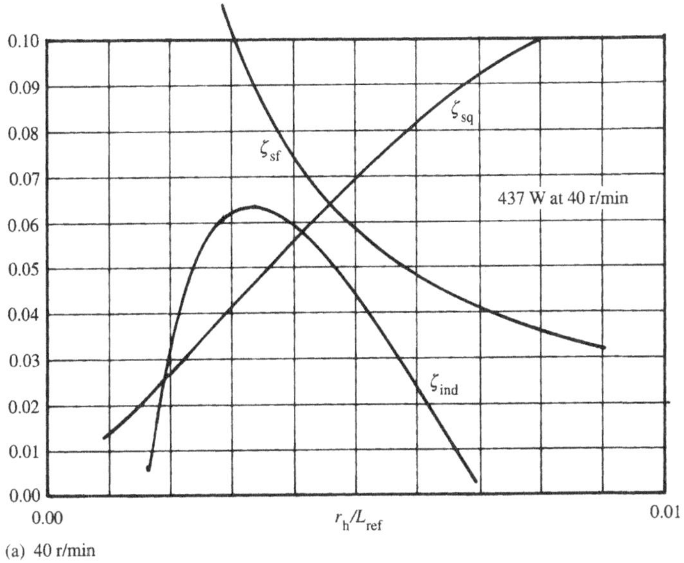

(a) Pumping loss, heat transfer loss, and net specific cycle work as functions of hydraulic radius at 40 r/min

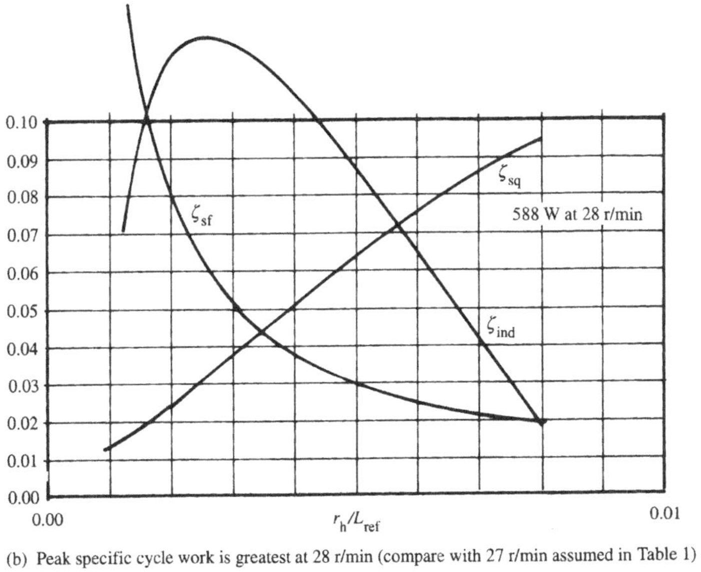

(b) The peak of specific cycle work is greatest at 28 r/min (cf. the value inferred for the provisional specification)

Fig. 12.2 Pumping loss, heat transfer loss, and net indicated specific cycle work as functions of hydraulic radius for two values of r/min

$$r_h = \frac{\P_v \{d_{warp} + 2d_{weft}\}}{\pi\{m_{warp} d_{warp} + 2m_{weft} d_{weft}\}} \tag{12.2}$$

For the special case where diameter and mesh of warp and weft are equal, equation (12.1) may be inverted to give the m_w corresponding to given volumetric porosity, $\P_v$, and r_h. For the values of 0.8 and 1.125 mm assumed at this point, m_w is 0.339, so that pitch = $1/m_w$ = 2.95 mm, or slightly less than ⅛ in. Equation (12.2) may then be inverted to give a numerical value of d_w. This is identical to r_h, namely, 1.125 mm.

However, the crucial insight from the diagrams is that, regardless of r/min, the range of values of hydraulic radius at which the engine produces power, i.e. the range of viable regenerator lay-ups, *is very narrow*. Had Stirling used half the r_h (e.g. by halving wire diameter, d_w, at the same volumetric porosity) *the engine would not have turned round* – even under no load.

There are three possibilities

(a) that Stirling was luckier in his regenerator design than anyone has been since;
(b) that he *designed* his regenerator, thermally and hydrodynamically, and did so very effectively;
(c) that operational success was achieved by a process of trial and error.

These possibilities are examined in turn.

When James Stirling refers to the prototype engine in his lecture to the Institution of Civil Engineers in 1853 he does not talk of endless rebuilds, or experimenting with the design of the regenerator to get it right. His precise words are *'it performed very well'*. The craft of wire drawing dates from ancient Egypt (Alexander and Brewer 1963). By 1818 wire was available in a wide range of sizes down to the fine diameters used for weaving sieves for paper-making. It is therefore unlikely that wire diameter was determined by availability. Conversely, the very range of available sizes minimizes the chance that the appropriate choice was coincidental.

An argument is going to be advanced in favour of the second scenario – that he performed calculations and carried out experiments to *optimize* the regenerator. It has taken 180 years, and much digital computing, to reach the point where heat transfer effectiveness can be balanced quantitatively against flow friction losses. The fact provokes a search for an explanation as to how the inventor might have achieved this in 1816. The simplified approach to optimization of Chapter 11 could well have been within Stirling's grasp.

12.3 Stirling's regenerator design

If Stirling is going to optimize the performance of his regenerator, he must work with two concepts

(a) the power consumed in pumping the air back and forth;
(b) the performance penalty (engine power equivalent) of incomplete temperature recovery.

For the former he needs to know how pressure drop relates to flow through his type of regenerator matrix. By 1816 flow resistance studies were far from new: the Frenchman Marriotte had carried out wind-tunnel tests in the mid-1600s (White, 1979). Stirling's physics studies would have covered Boyle's Law (1662) for gases and Bernoulli's studies (1738) on the relationship between pressure and speed in incompressible flow, viz, $\Delta p = -\frac{1}{2}\rho u^2$. Charles' Law (1787) was already long-established. A form of ideal gas law was in use (Fox 1986) by engineers debating the advantages of the expansive use of steam. Avogadro's new hypothesis (1811) would have been a hot topic among physicists in 1816, so Stirling would even have had an inkling of the molecular view of gases.

12.3.1 A suitable expression for pumping loss

He will know that pumping power in steady flow is the product of volume rate, Q', with pressure difference, Δp, viz. $Q'\Delta p$. He has not heard of Reynolds' number, N_{re}, but does not need to for two reasons: he is dealing with a single fluid (air), and at the high N_{re} of his regenerator (about 2000, see Table 11.1) flow is largely independent of N_{re}. A water U-tube manometer will serve to measure Δp. Bellows supply air at high volume rate and low pressure. (He has a smithy – and, from Monday to Saturday, the services of the church organ!) If he measures Δp against flowrate, Q', for a few different regenerator packings he will find $\Delta p/\frac{1}{2}\rho u^2 \approx -C_f L_r/r_h$. For his experiments (woven wires at $N_{re} \approx 2000$) C_f will be effectively constant at a numerical value of about 0.3. He now knows that

$$\textit{Pumping power} \approx \tfrac{1}{2}\rho u^2 Q' \frac{C_f}{r_h/L_r} \tag{12.3}$$

This is the inverse relationship between pumping power and his design parameter r_h/L_r already noted in the falling curve of Fig. 12.2. The physicist will *routinely* break down symbolic relationships into dimensionless parameters: this is how a limited range of experiments is exploited in support

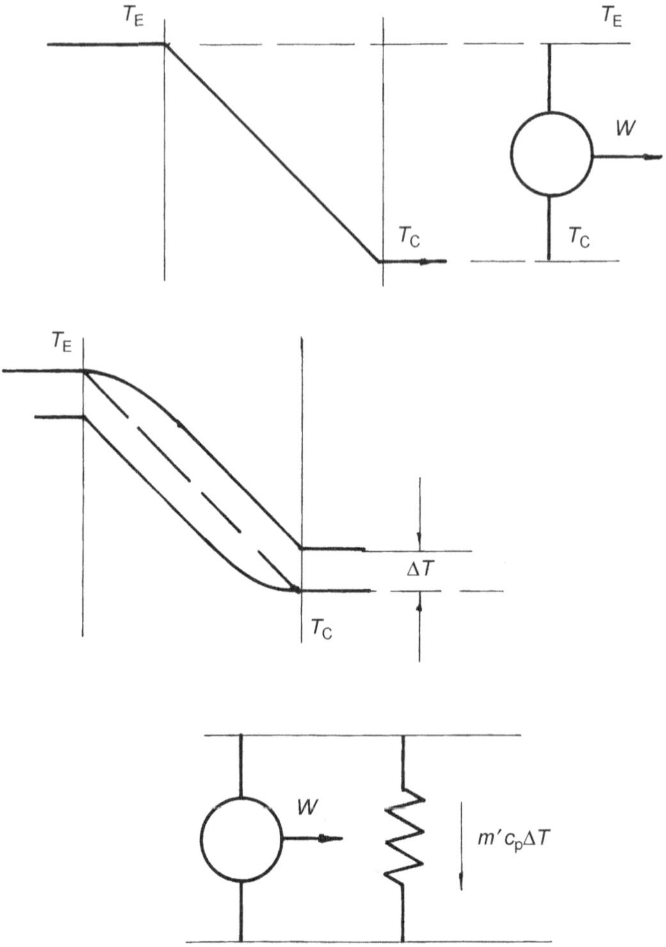

Fig. 12.3 Schematic comparison of ideal (lossless) regenerator temperature distribution and realistic distribution involving heat exchange across temperature difference ΔT

of a law of wide generality.

He now needs an expression for the second curve of Fig. 12.2, the loss corresponding to incomplete temperature recovery which increases with increasing r_h/L_r. Figure 12.3 compares temperature profiles for: (a) 100 per cent temperature recovery, and (b) for reduced recovery ratio. The latter case may be thought of as resulting in a flow of *excess enthalpy*, H', at regenerator exit equal to $m'c_p\Delta T$, where m' is mass rate, c_p is specific heat (Stirling knows all about c_p), and ΔT is the difference between actual air outlet temperature and the ideal value at 100 per cent recovery ratio. How H' becomes a performance penalty will be addressed shortly. Stirling can make use of it only if he can quantify ΔT. If the usual chronology of the evolution of regenerator theory is accepted, this is what he lacks the means to do.

If there is difficulty in accepting that Stirling might have tackled this problem analytically, and indeed tackled it with success and ahead of Hausen, the difficulty persists only as long as he continues to be seen as a village pastor with a sideline interest in technology. He was not; he was physicist, engineer and savant – and exceptionally gifted in all three roles. It is worth considering also that had a head start on a number of counts: he would not, for example, have been aware of the distractions the mathematicians were to inflict on regenerator theory. His patent description is evidence of prior analytical work along appropriate lines.

(a) It confirms that he visualized operation in terms of temperature *difference*, ΔT, rather than of the irrelevant absolute temperature;
(b) It gives a correct interpretation of the influence of changing specific heat;
(c) It correctly identifies the ratio (passage length/hydraulic radius, r_h/L_r) as a crucial performance parameter;
(d) It states *in words of one syllable* the condition for operation at maximum efficiency which took Hausen much symbolic manipulation to deduce 120 years on.

12.3.2 The temperature solutions

A scientist of Stirling's calibre is capable of breaking the temperature analysis down into four simple stages:

(1) He will note the high thermal capacity ratio: in the context (engine unpressurized) the ratio of the thermal capacity of the wire matrix to that of the gas per blow was massive – of the order of 10 000. This fact permits temperature fluctuations of the wires to be dismissed. In these special circumstances, analysis simplifies even beyond the dramatic gains of the formulation of Chapter 11.
(2) He will follow a particle entering from one end of the regenerator (as before) and note that temperature difference, ΔT, between particle and matrix (which he needs to evaluate) increases at the rate $-\underline{u}\mathrm{d}T_w/\mathrm{d}x = -\underline{u}(T_C - T_E)/L_r$, where $\underline{u}$ is the speed of the particle.
(3) He knows about Newton's Law of Cooling

$$mc\ \mathrm{d}T/\mathrm{d}t = -hA_w\ \Delta T \qquad (12.4)$$

and is in a position to use the following reasoning as an alternative to the intuitive development of Chapter 11: the m of equation (12.4) is the $\rho A_{ff}dx$ of the gas 'particle' of Fig. 11.3b; T is its temperature, T_g; the c of equation (12.4) is c_p for the gas.

The final step is lengthier but no less simple. The symbol for rate of change of T_g, namely, dT/dt is rewritten DT/dt consistent with the fact that it is being applied to the moving 'particle'. Substituting into equation (12.4) gives an equation relating T_g on the left-hand side to ΔT (= $T_g - T_w$) on the right. This will not do as it stands, but may be dealt with via a standard analytical trick of subtracting DT_w/dt from both left- and right-hand sides:

$$\rho A_{ff}\, dxc_p\, D/dt\{T_g - T_w\} = \rho A_{ff}\, dxc_p\, D\Delta T/dt = hA_w\, \Delta T - \rho A_{ff}\, dxc_p\, DT_w/dt$$

The definition of D/dt (the 'substantial' derivative) is well-known to Stirling the mathematician as $\partial/dt + u\partial/dx$. He now has an equation in ΔT. The first term relates to variation with time at constant location, which is zero for T_w (because of the near-infinite thermal capacity ratio), and so drops out. This leaves a right-hand side consisting of the terms $hA_w\Delta T - \rho A_{ff}\, dxc_p u\partial T_w/\partial x$. Substituting for $\partial T_w/\partial x$ and rearranging leads again to the equation arrived at independently in Chapter 11

$$D\Delta T/dt = -\underline{u}\partial T_w/\partial x - |\underline{u}|N_{st}\Delta T/r_h \qquad (12.5)$$

As before, the differential coefficient, $D\Delta T/dt$, is zero for the part of the solution he is interested in (temperature difference at exit), leaving an expression for ΔT in terms of NTU – i.e. of N_{st} combined with his geometric design parameter r_h/L_r:

$$\Delta T_{exit} \approx \frac{-(T_C - T_E)}{NTU} \approx \frac{-(T_C - T_E)r_h/L_r}{N_{st}} \quad NTU >> 1 \qquad (12.6)$$

For the conditions of Stirling's experiments, Stanton number N_{st} will not vary greatly from a value of 0.03, and he may choose to regard it as a constant. This makes ΔT proportional to his geometric design parameter, r_h/L_r (recall Fig. 12.2). It remains for him to quantify the performance penalty of ΔT.

Engineers of the era were routinely constructing indicator (p–V) diagrams (Fox 1986) in connection with the debate over the benefits of expansive use of steam. In doing so they grappled with the added complication of an equation of state for steam. Stirling, had he chosen, could readily have formulated an expression for the indicated power of his engine in terms of operating temperatures, T_E and T_C. Had he done so, he may well have worked in terms of the overall difference, $T_E - T_C$, since he not only cites a specific value ($T_E - T_C = 480$ °F) for rated operation, but is categorical about the modifications to volume ratio required when the difference exceeds or falls short of this value.

To suggest how he approached the matter of power calculation would be to take speculation too far. On the other hand, the analytical treatment of *any* expression to quantify the effect of varying temperature recovery deficit, ΔT, is routine. An illustration by reference to the Schmidt analysis thus covers a wide range of Stirling's options.

Ideal, indicated power is simply $P_{Sch} = \eta_{Sch} Q_E'$, in which η_{Sch} is Schmidt thermal efficiency, $(T_E - T_C)/T_E$, and Q_E' is rate of heat addition according to the Schmidt algebra. Stirling does not yet know about thermal efficiency; Joule does not establish the mechanical equivalent of heat until 1843, and 'efficiency' is still in terms of pounds of coal per hour per horsepower. However, he does not need to: $(T_E - T_C)/T_E$ is a factor of any algebraic cycle treatment based on cycle temperature limits and a gas law relating p, ρ, and T. In terms of overall temperature *difference* it is $[T_E - T_C]/([T_E - T_C] + T_C)$, in which the 'spare' T_C is essentially a constant. The Q_E' term in standard formulation is also a function of temperature ratio, $N_T = T_E/T_C$, and may thus be rewritten in terms of $[T_E - T_C]$. As it is a weak function the refinement will not be applied here.

The idealized effect on P_{Sch} of changes in $[T_E - T_C]$ is given by differentiation:

$$\frac{dP_{Sch}}{d[T_E - T_C]} = \frac{T_C Q_E'}{\{[T_E - T_C] + T_C\}^2} \tag{12.7}$$

The highly idealized Q_E' will be a gross overestimate of actual indicated heat input, but Stirling is more than an analyst: if he expresses equation (12.7) as a *fractional* dependence (by dividing by P_{Sch}) he eliminates this aspect of the idealization altogether!

$$\frac{dP_{Sch}}{P_{Sch}} = \frac{T_C d[T_E - T_C]}{[T_E - T_C]\{[T_E - T_C] + T_C\}}$$

From Fig. 12.3 the numerical value of $d[T_E - T_C]$ is simply $2\Delta T$, which is available to Stirling in terms of NTU and thus in terms of his geometric design parameter r_h/L_r.

$$\frac{dP_{Sch}}{P_{Sch}} = \frac{dW_{Sch}}{W_{Sch}} = \frac{2T_C r_h / L_r}{N_{st}\{[T_E - T_C] + T_C\}} \tag{12.8}$$

Stirling has the option of differentiating the sum of his losses in analytical form [equation (12.1) + equation (12.8) duly reduced to a common denominator of his choice] with respect to L_r/r_h, setting equal to zero, and extracting an explicit expression for the optimum. Chapter 11 has illustrated the procedure. Alternatively, he can plot diagrams like Figs 11.7 or 12.2. Either method, graphical or analytical, applied to the 1818 design produces the same r_h/L_r.

It is time to consider the third scenario – trial and error.

12.4 The alternative

For this time-honoured process to have been the route to success in 1818 it is necessary to imagine the fire being lit for the first time, and the engine failing to function – or functioning at a fraction of the output hoped for. Stirling's interpretation must then be that the fault lies with the regenerator rather than with crank kinematics, or with operating temperature, or with friction or leakage. In other words, he must be convinced that the problem lies with regenerator lay-up rather than with the phase angle, volume ratio, hot-end temperature, or friction.

In order to remedy, the cylinder has to be dismantled from the masonry. This calls for the use of lifting tackle. The displacer, measuring 2 ft diameter by 4 ft long, has to be removed, the original regenerator unwound and a replacement woven into position using a different diameter of wire and/or winding pitch.

If we have deduced correctly that a wire diameter close to 3/64 in is essential for satisfactory operation, the *length* of wire required to produce a matrix of 75 per cent volumetric porosity is 4.133 miles. The number of possible combinations of wire diameter and winding pitch is infinite, and the processes of dismantling, rewinding, reassembly, and firing have to be repeated in the expectation that a workable combination will be hit upon. Finally, the time required for the complete cut-and-try approach has to be reconciled with the short period (1816–1818) between design and operational service.

A development programme of such fundamental consequence to the concept would surely have been a matter for report in brother James' accounts to the Institution. By constrast, he reports that it worked – though '*... not to the power expected*'. Expected? What expectation? Whose expectation? This is an engine without precedent for any aspect of its functioning. How could there have been expectation other than the designer's calculated power output.

12.5 Résumé

Where have we come to in this assessment of Robert Stirling?

(1) He defined practical quasi-reversible heat exchange 8 years ahead of Carnot's famous treatise.

(2) He conceived, designed, patented, built and put to work an engine with no history of modification or development – an engine which distinguished itself pumping water from a quarry.

(3) The achievement may be contrasted with that of Carnot – who did not design an engine, did not patent, and did not build. The Carnot engine is a 'thought experiment' only. Carnot himself did not get the Second Law of Thermodynamics completely right (Spalding and Cole 1966), yet he is credited with the modern concept in every text on basic physics and engineering, and is celebrated accordingly. The Scots minister beat him to it by 8 years, but goes relatively unsung. He even has to endure the indignity of having his thermodynamic cycle incorrectly explained in the limited number of engineering texts which concede a mention.

The final chapter which follows identifies potential in the 1818 design as an alternative to the pressurization which has become the norm where high specific power is required. It also proposes measures to give long overdue recognition to the scientific genius and engineering prowess of the inventor.

13

. . . . . and the heyday to come

13.1 Full circle?

By comparison with the steam engine, two striking features of Stirling's engine were simplicity and safety. Statistics for the era are lacking, but it is recorded (Percival *et al.* 1960) that in the 17-year interval between 1862 and 1879 over 10 000 boiler explosions with resulting death and injury occurred in England alone, i.e. a rate of about 10/week! That the industrial air engine persisted was partly a response to this very problem. Late-twentieth-century Stirling engines now demonstrate specific power and efficiency which the inventor could scarcely have foreseen, but at a cost in terms of sophistication out of line with commercial reality. The emphasis, almost two centuries later, is again on simplicity, with the promising results recently demonstrated by the Viebach engine.

13.2 An air engine to challenge hydrogen and helium – the Viebach CHP unit

If the Stirling engine has a commercial future in the short term, it appears at the time of writing to be tied to that of domestic combined heat and power (CHP). A number of designs are under development for the purpose: kinematic and free-piston, N_2-charged and helium-charged (Meek, 2000). Into this arena has stepped Stefan Viebach, who in a single design iteration has achieved gas-to-electrical conversion efficiency to match those of the long-established players, together with exhaust emission levels to surpass them. An illustration of the combined engine–generator is not available, but Fig. 13.1 shows its immediate predecessor, made by Stefan's father, Dieter Viebach.

The significance of the achievement for the present account is that it is the result of ruthless application of the design priorities appropriate to the application: a matching of concept to context. First and foremost, combustion chamber problems have been forestalled by designing the engine around a developed, proprietary FLOX (FLameless OXidation) chamber, rather than designing a combustion chamber around the engine. Potential sealing and containment problems have been minimized by specifying nitrogen as the working fluid. Choice of the 'V'-configuration 'alpha' layout permits use of the simplest possible drive mechanism – a single crankshaft with two connecting rods on a common crank-pin. Gas circuit proportions have been calculated with

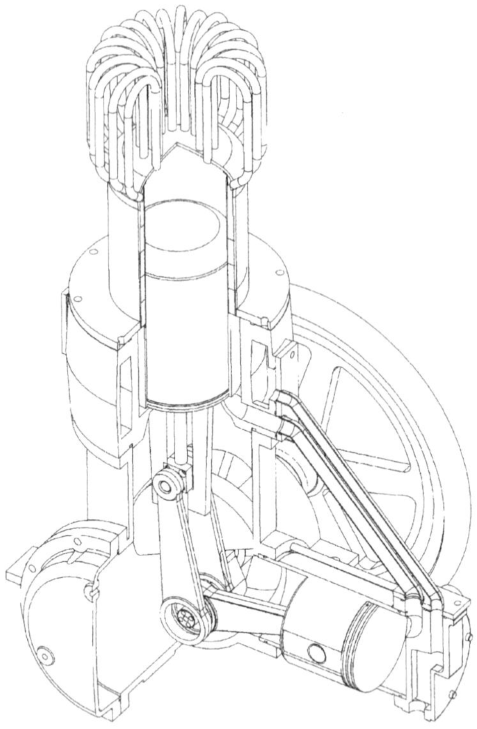

Fig. 13.1 The nitrogen-charged engine from which the Viebach CHP system has been developed in a single design iteration

the collaboration of Peter Maeckel of Kassel University using the scaling methods outlined in Chapter 10. Rotating bearings are of sealed rolling-element type sized in terms of load and r/min for rated life. As many components as possible are simple castings in aluminium alloy. The electric generator is sealed inside the pressurized crank-case and serves as a starter. The start cycle is automated, and operation is initiated at the turn of a switch. Further details are given in Table 13.1

13.3 A bold initiative from New Zealand

Meek's report for the Guardian (*op. cit.*) centres on two Stirling CHP systems. The first is a development by British Gas of the Sunpower free-piston concept. The second is a unit designed by Clucas and Raine (1994) originally as a battery charger for marine use, and currently in batch production by WhisperTech of New Zealand. According to EA Technology who are championing application of the WhisperTech engine to domestic CHP, the

Table 13.1 Outline specification of the Viebach Stirling-powered CHP system

Construction	Simple Stirling engine of α configuration: V-90° crank drive, rolling bearings lubricated for long life, dry-running sliding bearings. Pressurized, hermetically sealed crank-case. Engine entirely out of castings in aluminium alloy and grey iron. Expansion cylinder precision casting in alloy steel
Generator	Integral flywheel generator, asynchronous generator for parallel grid operation. Maximum generator efficiency 92%
Displacement	2 × 520 cc
Bore	105 mm
Stroke	60 mm
Mean gas pressure	10 – 25 bar
Working gas	N_2
Heater temperature	750 – 800 °C
Upper gas temperature	700 °C in heater tubes
Lower gas temperature	90 °C in cooler channels
Cooling water temperature	< 40 °C at inlet < 70 °C at outlet
Drive r/min	750 r/min parallel grid drive with asynchronous generator
Thermal output	3 – 9 kW, peak burner load 20 kW
Electrical capacity	Design: 2.5 kW at 750 r/min Measured: 2.1 kW at 800 r/min with synchronous generator
System efficiency	Design: $P_{electrical}/P_{gas\ burner} > 28\%$ Measured: $P_{electrical}/P_{gas\ burner}$ 22%
Burner	Gas burner for flameless oxidation with integrated preheater from WS Waermeprozesstechnik GmbH, Renningen
Exhaust gas analysis	Nox < 10 mg/m^3 (own measurement) CO << 10 p.p.m. (own measurement)
Life between major overhauls	80 000 h

technical barriers are overcome to the extent that *'it's simply a matter of. . . things like designing the right colour for the casing'*.

Figure 13.2 is a simplified cross-section showing a four-cylinder 'Rinia' configuration driven by a nutator mechanism or 'wobble-yoke'. The electrical generator is hermetically sealed within a crank-case pressurized with nitrogen to a maximum of 22 bar. The annular regenerators are wound 'bandage' fashion from woven wire mesh.

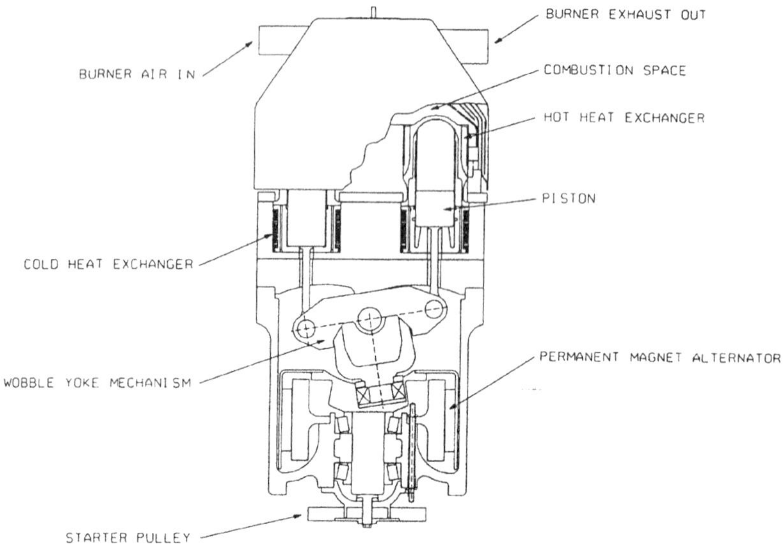

Fig. 13.2 Four-cylinder, Rinia-configuration N_2-charged engine in batch production by WhisperTech of New Zealand

If the 1994 account accurately describes the production version, expansion exchanger design derives from experiments affording a lesson for thermodynamic design: brake power yielded by a slotted exchanger (cf. the Philips MP1002CA) was plotted as a function of r/min. The exchanger was then substituted by a series of units having the same external slots, but smooth bores affording a range of annular gaps. A gap of 0.6 mm gave a brake output only about 10 per cent down on that of the original slotted exchanger – and a considerably more economical construction. The finding supports an expectation that, where the performance criterion is power, heat exchange area may be traded for dead space and concomitant increased compression ratio. Matters of interest not disclosed by the account (Clucas and Raine *op. cit.*) are: (a) whether the benchmark slotted exchanger was itself an optimum, and (b) the effect of the simplification on thermal efficiency.

Electrical output at 12 or 24 V d.c. is 850 W. Usable heat to water in CHP mode is 6 kW, with an overall energy recovery of about 90 per cent. Orders at August 2000 stood at 120 units (Clucas 2000). This is within striking distance of Philips' batch-production output of 150 MP1002CA sets, and is no doubt a factor in EA Technology's optimistic view of prospects.

13.4 Future of the 1818 concept

The revival of interest in the Stirling engine in the 1930s gave long-overdue recognition to the importance of the regenerator, but sought high specific power through pressurization and the replacement of air by lighter gases. A case for reconsideration of air (or N_2) has already been made, and reopens prospects for unpressurized types.

Pressurization increases the mass of working fluid processed per cycle. If all things remain equal, cycle work increases in proportion. They do not, however. At given cyclic speed, local, instantaneous N_{st} decreases. Maintaining *NTU* calls for increased heat transfer area. In the modern Stirling engine with tubular heat exchangers, increased area generally means increased dead volume. This reduces pressure ratio, calling for increased charge pressure, increased heat transfer area, and so on.

With heat rate proportional to the product of N_{st} with temperature difference, ΔT, the alternative approach is to maintain N_{st} high (low density, low charge pressure) and increase ΔT by increasing pressure swing. The result is the high-compression-ratio engine. The performance of the 1818 engine was inherently limited by the massive hydraulic radius of the variable-volume spaces and the correspondingly low *NTU*. On the other hand the coaxial configuration in conjunction with the annular regenerator are a first condition for achievement of high pressure ratio.

Two contrasting approaches to Stirling engine design emerge: (a) high dead space, high charge pressure, low compression ratio versus (b) low dead space, high compression ratio, unpressurized. The former category is exemplified by the Viebach engine. The latter derives directly from the original concept. In principle, it avoids many of the manufacturing challenges of the pressurized derivatives. A commercial application, however modest, would amount to a full-circle return to every aspect of the inventor's original concept – a true Stirling engine in the fullest sense.

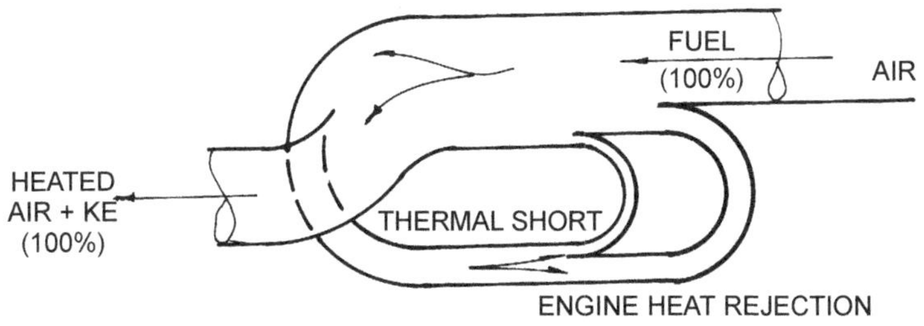

Fig. 13.3 Sankey diagram of gas-powered hair drier concept, CHP in miniature!

The smaller the engine, the greater the opportunity for exploiting the feature. The next section describes a tailor-made application, and in sufficient detail to allow evaluation of commercial prospects.

13.5 A gas-powered, cordless hair drier?

In early 1997 an adviser to Messrs Cambridge Consultants Ltd (CCL) requested a view as to whether a small Stirling engine could power the air impeller for a portable, gas-powered hair drier. As a concept there were immediate attractions.

(1) Here is CHP (combined heat and power) in miniature. CHP permits relaxation of demands on combustion system efficiency, since heat not converted by the prime mover is exploited elsewhere.

(2) In contrast with the power required to heat the air stream (up to 1.6 kW in a domestic hair drier) fan power is a few Watts. Model-makers have been known to claim brake powers of 0.5 W/cm^3 swept volume for unpressurized 'hot-air' engines. On the basis that no hot-air engine known to the author has ever had the benefit of thermodynamic design, a four-fold improvement in specific power might be a realistic expectation of a properly designed engine equipped with a regenerator – or some 30 W for 15 cm^3 swept volume.

The concept extends without essential modification to hot-air guns for paint stripping and shrink fitting, thereby broadening commercial prospects.

If the proposition sounds fanciful it may be worth considering that the Toshiba Corporation has toyed with the idea (see below). Moreover, sophisticated styling combs powered by gas from replaceable plastic cartridges are already widely marketed by Braun. Judging by displays of replacement cartridges at stockists, they are popular.

High engine efficiency is clearly not a requirement, but it may be suspected that the ratio of mechanical to thermal energy will become a consideration as the concept evolves. A Sankey diagram (Fig. 13.3) indicates the relevant energy flows in both sense and magnitude, the latter by the relative width of the flow channel. With T_∞ as enthalpy datum, ambient air enters with zero energy. The engine-driven fan adds kinetic energy subsequently supplemented by heat rejected from the engine cycle via cooling fins. The flow of slightly pre-heated air divides into combustion and bypass streams. The combustion stream is heated to high temperature in the vicinity of the engine cylinder head by reaction with the fuel supply, and maintains engine expansion exchanger at operating temperature. The exhaust stream mixes with the bypass air, raising the combined temperature to T_{exit}. For all but unrealistically low thermal

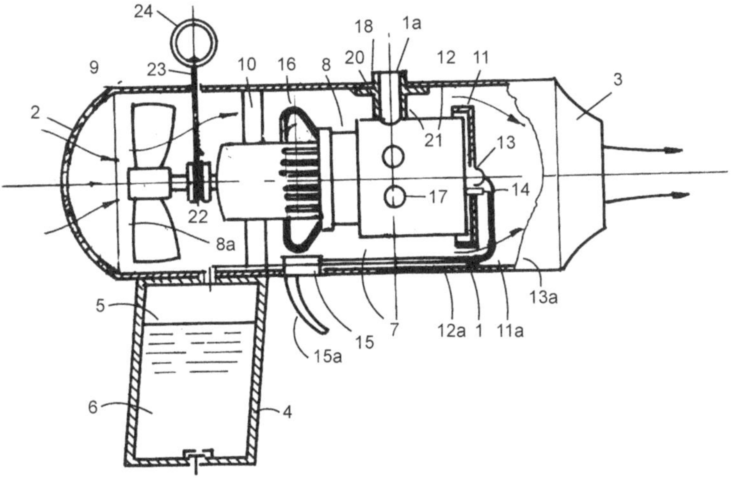

Fig. 13.4 Drawing accompanying the abstract of the Toshiba patent

efficiencies, excess fuel is burned to maintain rated T_{exit}. In the Sankey diagram of Fig. 13.3 the recirculated energy is shown to exaggerated scale in anticipation that actual values will be small fractions of total energy throughput.

Prior disclosure of the basic concept by the Toshiba Corporation pre-empted patent cover. Moreover, it has since been discovered (Feulner 1999) that a German patent exists. Lack of effective proprietary protection in turn precluded the chance of significant funding. As an academic design exercise, on the other hand, here is a project which challenges precisely the appropriate skills and resources. To the time of writing, detail design has been taken as far as machining the basic components. Operational success thus remains to be demonstrated, but presentation of this account may be justified to the extent that it advances the concept way beyond the Toshiba disclosure, the technical status of which may be judged from the patent drawing reproduced in Fig. 13.4. In combination with the published abstract, it is subject to the following misgivings.

(1) It is not based on a functioning prototype: the crankshaft is shown emerging coaxially with the cylinders, suggesting a drive mechanism of excessive complexity (or of previously unimagined ingenuity); impeller diameter is inadequate, as is the free-flow area of the apertures admitting the air flow; an unpressurized tank for liquid fuel is implied with no evident means of getting the fuel to the vaporizer; the proposed starting means promises a substantial bending moment on the (overhung) drive shaft.

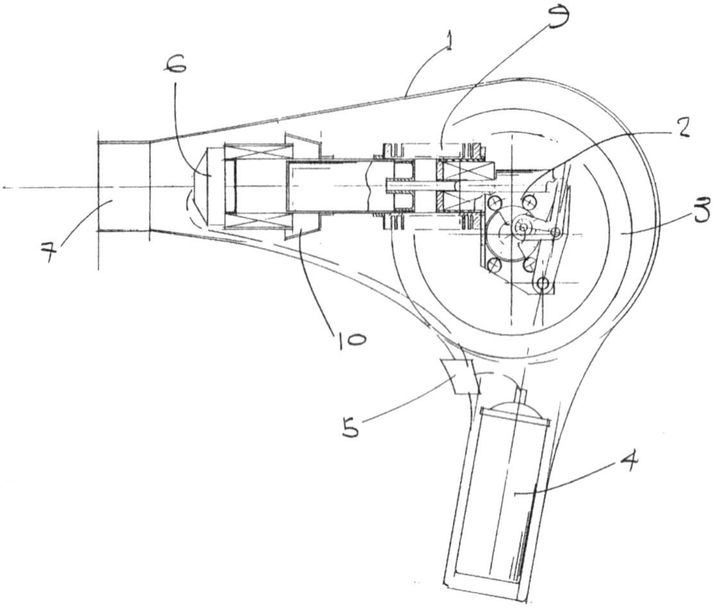

Fig. 13.5 The drier concept as an application of the steady-flow energy equation

(2) The combustion chamber exhausts sideways through the drier, heat to the drying air stream thus being by conduction followed by convection and radiation from metal surfaces of limited area. This contrasts with the unlimited heat transfer available by direct heating of the air stream

The patent abstract discloses nothing of technical value, and design of the competing system must start from scratch. Engine power and speed, for example, must be set in terms of air flow requirements. Measurements from a mains electric drier suggest an electrical input to the fan of 15–20 W, only part of which, of course, will be imparted to the air stream. The power estimate may be corroborated by application of the celebrated steady-flow energy equation. This relates outlet temperature, T_{exit} (Fig. 13.5), mean outlet velocity, $\underline{u}_{exit}$, outlet orifice area, A_{ff}, air density at outlet, ρ_{exit}, and power input. T_{exit} is taken as 100 °C and inlet conditions as T_∞ = 20 °C and $\underline{u}_\infty = 0$. Outlet velocity now follows by use of outlet orifice area, A_{ff}, of the conventional mains electric unit:

$$q' + W'_f = \rho_{exit}\underline{u}_{exit} A_{ff}\{\tfrac{1}{2}\underline{u}_{exit}^2 + c_p(T_{exit} - T_\infty)\}$$

Inclusion of the kinetic energy term makes this a cubic equation in the unknown mean mass velocity $\underline{u}_{exit}$ where

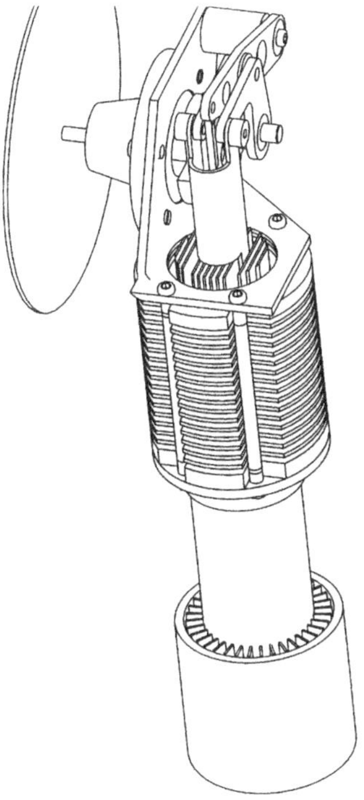

Fig. 13.6 Pro-Engineer rendering by Oon (1998) of the high-compression-ratio air engine

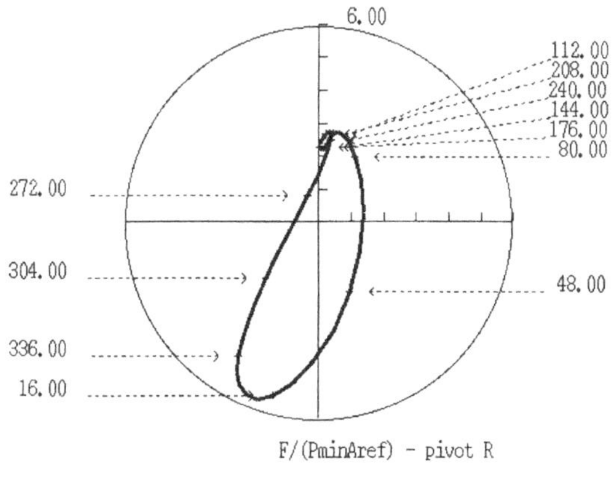

Fig. 13.7 Cycle simulation integrated with crank kinematics produces polar diagram of loads on all rotating bearings. The specimen here is for the inboard crankshaft journal bearing

$$\underline{u}_{exit}{}^3 + 2c_p(T_{exit} - T_\infty)\underline{u}_{exit} - 2(q' + W'_f)/\rho_{exit}A_{ff} = 0 \qquad (13.1)$$

Tentative use of 12 W for W'_f (power *actually imparted to the air stream*) and evaluation of the coefficients and solution of the cubic by the standard formula (Press *et al.* 1989) gives $\underline{u}_{exit}$ = 11.36 m/s. Volume flowrate, $Q' = \underline{u}_{exit}A_{ff}$ = 0.0181 m^3/s.

Assuming that a fan power achieving Q' with an ambient temperature flow will achieve the same Q' hot avoids problems with small numerical differences. Equating the mechanical input rate of the fan, W'_f, to the increase in kinetic energy of the stream, $\frac{1}{2}A_{ff}\rho\underline{u}^3$, but provisionally ignoring flow friction through the system, gives W'_f = 15.89 – or, say, 16 W.

This is an obvious trial application of the high-compression-ratio engine with annular regenerator. Figure 13.6 is a Pro-Engineer rendering by Oon (1998) of a version similar in all respects the to 1818 engine, except for a recently patented crank mechanism – and, of course, for the 15 cm^3 swept volume. The price usually paid for the advantageous compression ratio of the coaxial layout is complexity of the crank mechanism, which must actuate piston and displacer along a common axis and with minimum side thrust. The new mechanism not only minimizes side thrust on piston and displacer, but affords considerable kinematic flexibility. The offset of the crankshaft centre-line from the main cylinder axis happens to fit particularly well into the envelope of the drier.

The computer simulation embodies precisely the gas-circuit model developed for the study of the 1818 engine, but substitutes the kinematics of the new drive mechanism. In addition, loads and corresponding friction losses at all bearing points are calculated as a function of crank angle. Figure 13.7 shows inboard main bearing load in polar diagram representation. Friction effects corresponding to instantaneous loads at all contact points are summed over a cycle in the process of computing mechanical efficiency.

Formulating the simulation as a function of the drive kinematics and subjecting to an optimization algorithm automatically exploits the potential for high compression ratio and correspondingly enhanced specific power. A conservative interpretation of simulation results suggests 2 W/cm^3 or 30 W brake output at 4 000 r/min. While still an order of magnitude below the comparable internal combustion engine, performance is set to exceed that of the toy 'hot-air' engine traditionally taken to represent the potential of the unpressurized type. More importantly, it promises to be sufficient to power the drier.

Figure 13.8 shows a selection of completed components. The regenerator consists of annuli cut from Retimet™ metal foam sheet on a numerically controlled wire erosion machine. The elements are installed end-to-end over the machined ceramic displacer body to give the required stack length. The

Fig. 13.8 Selection of components of hair drier engine

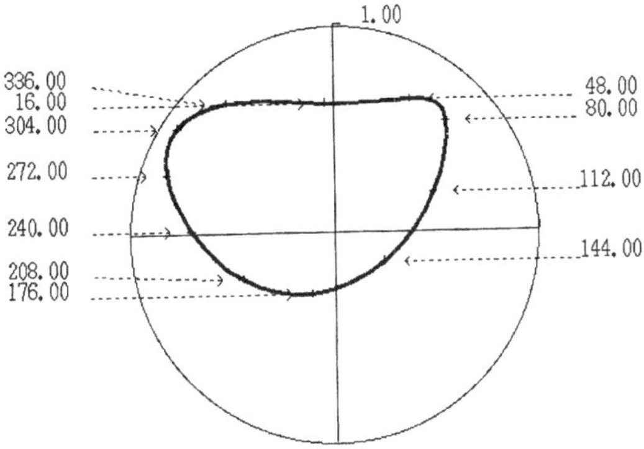

Fig. 13.9 Polar diagram of out-of-balance load referred to crankshaft centre-line

Retimet grade selected is that having hydraulic radius closest to the value indicated by the optimization studies. The aluminium alloy heat sink is wire cut to give cooling fins aligned with the axis – and thus with the direction of air flow through the drier.

A hand-held appliance calls for dynamic out-of-balance to be minimized. Dedicated software calculates net out-of-balance force of all moving components, refers it to the crankshaft axis and displays it in the form of a polar diagram. Figure 13.9 is a specimen. The optimum balance mass moment, $m_b r_b$ [kg m], to be applied to the crankshaft is that which minimizes the envelope of the polar trace. The value of $m_b r_b$ determined in this way can be incorporated with some precision by cutting away the crank-web to form 'ears' to an exact geometric formula (Organ 1990).

The traditional air-cooled toy 'hot-air' engine suffers from two 'thermal bottlenecks' which can be expected to restrict performance. The present design tackles both. The first bottleneck is due to the fact that the heating flame normally plays on the end of the expansion cylinder before exhausting to atmosphere. By contrast, the present design provides for a two-stage process of convective heat transfer: after impinging on the cylinder end, the combustion products pass through radial holes formed in an extension integral with the cylinder. From there they enter a cylindrical sleeve of machinable ceramic to be guided through radial fins brazed to the outside of the cylinder. A second bottleneck occurs when the heat from the cycle has to be rejected, together with the heat of thermal shorting, via fins external to the compression cylinder. In the present design, the absence of piston side-thrust offers the possibility of a skirtless piston. This can be carried on a seal ring and a light bearer ring. Fins substitute the skirt and protrude into the air stream, dissipating heat from the piston crown directly. The piston motion allows and encourages contact between the external cooling air, fins and the *inside* of that part of the compression cylinder also wetted for part of the cycle by the working fluid.

With limited power available from the engine, impeller design is crucial. The relatively low r/min calls for a larger diameter than that of the mains electrical drier, and a return to the envelope shape of electrical models of the 1960s. If the diameter of the fan casing is not to increase yet further, an axial fan is the sole option. It was thus fortuitous that the evaluation of the drier proposal coincided with a study of Larrabee's analysis of the minimum-induced-loss propeller.

An axial fan is a propeller with multiple blades. The only objection to immediate use for impeller design is that operation in the hair drier fan probably comes closer to the 'static thrust' case than to in-flight operation of the Larrabee formulation, and that the original analysis is singular for that case (airspeed, V_α, = 0). The problem is algebraic rather than aerodynamic, and arises out of the use of V_α as normalizing parameter. The difficulty calls for complete reformulation which, however, has been accomplished without compromising any of the unique elegance of the Larrabee treatment. The task of reformulation, undertaken specifically in the interests of the study of the drier, was not trivial, and has been written up in the form of a paper (Organ 2000d).

Proprietary gas cartridges contain some 14 g of liquid gas, and at 28 mm diameter and 86 mm long are a convenient size for the proposed drier. In the styler application a cartridge lasts some 7 h. Maximum duty cycle time of the proposed drier variant embodying the same cartridge is estimated from power (1.6 kW), fuel calorific value (45×10^6 J/kg for typical hydrocarbon fuel) and fuel mass as 6.5 min. Longer drying periods would call for an enlarged fuel cartridge – and for associated development and launch costs.

13.6 A shot in the dark

The hair drier concept does nothing to remove the 'technology mismatch' inherent in the simple unpressurized air engine. Rather it is a chance application where the mismatch is turned to advantage. To make a widespread reality of the cheap, compact, quiet, externally heated prime mover in the fractional-kW range what is required is a means, *any* means, of drastically reducing the thermal bottlenecks so that the work-producing potential of the increased compression ratio can be realized.

Provided that we are prepared to aim for improved specific power and accept such thermal efficiency as results, the solution may be easier than expected. There is, in fact, only one thermal bottleneck – at the compression end, where the temperature difference driving convective heat exchange is limited to that set by ambient temperature. Expansion end temperature, by contrast, can be increased to the limit set by the creep-to-rupture strength of the latest materials. At the modest pressures involved an increase in T_E of 100 or even 200 °C above the 600 °C should not be a problem.

So, what of the compression end? Well, why confine cooling of the air charge to convection? Why not exhaust and scavenge it in the fashion of the simple two-stroke internal combustion engine and replace it with a charge at atmospheric pressure via the crank-case?

The moment the heat transfer bottleneck yields, a case arises for compression ratio to be the highest possible. Values in excess of 10:1 may sound ambitious, but it is worth noting there are no problems of pre-ignition or detonation as in the petrol engine, so the limit is set only by unswept volume at the piston inner dead-centre position.

Figure 13.10 is an orthographic view projected from a three-dimensional model of the engine built in Pro-Engineer and animated to check kinematic functioning. (The section-lining convention would not have been the first choice of the author.) Operation is inverted to permit heating by a natural convection flame. Bore, strokes and rated r/min are identical to those of the Philips MP1002CA air engine (Fig. 5.4) so as to afford a yardstick. The modest r/min allows the two-stroke porting to be shallow. This in turn means that the

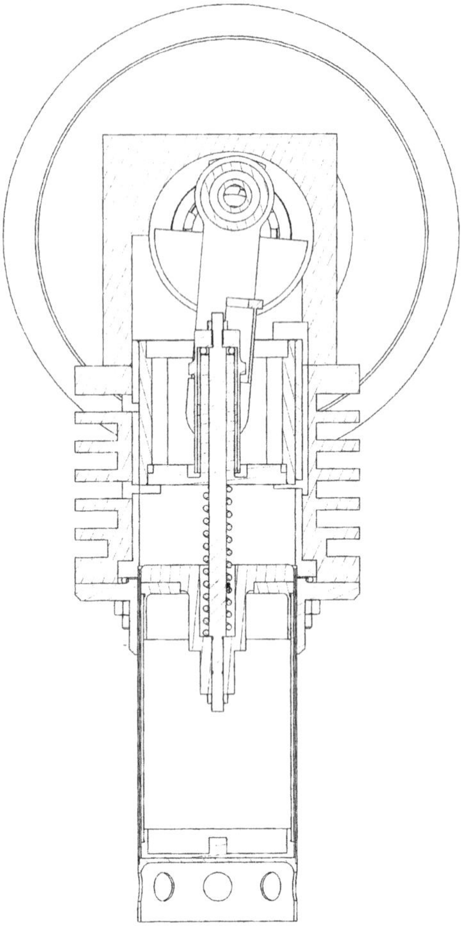

Fig. 13.10 The porting system of the two-stroke internal combustion engine enhances heat rejection capacity at the compression end, and encourages an investigation into the potential of ultra-high compression ratio

majority of the stroke is available for useful work.

A patent has been applied for to cover the combination of porting and displacer drive. The latter is best regarded as a vehicle for exploring the benefits of the former, the inventor being under no illusions about the potential of this particular drive for extended service. The essence is a spring which holds the displacer at its inner- (lower) dead-centre position during the compression stroke. A narrow annular gap around the displacer provides an element of thermal regeneration while keeping unswept volume to a minimum.

The uppermost extremity of the displacer drive rod is rigidly connected to a stepped sleeve guided on dry-running (Glacier DU) bushes on the central boss of the piston. A latch pivots on the two overhung gudgeon pins. While the connecting rod lies to the right of the cylinder axis on the compression stroke the latch is held against the side of the rod by light springs (not illustrated). As the rod swings through the vertical at inner- (lower) dead-centre, the latch engages below the step on the sleeve. The displacer, with the main spring now fully compressed, travels held against the piston to the common outer dead-centre position (not quite the uppermost position of the crank-pin because of the désaxé offset). As the connecting rod swings past the vertical it contacts 'ears' on the latch and releases it from the stepped sleeve. The spring drives the displacer downwards until it is arrested by a resilient stop between the stepped sleeve and the central piston boss. There is thus no impact between displacer crown and cylinder head. The latter is then delivered gently against the former by the initial, low-speed downwards motion of the piston. The displacer motion under the action of the spring does not have to be particularly rapid, since there is no point in its reaching inner dead-centre before the piston has closed the exhaust port.

The drawing shows the transfer port and the provision for under-the-piston induction of atmospheric air. In extended operation the latter would be filtered. The air exhausts at a temperature above atmospheric and, unlike the exhaust from the internal combustion engine, is clean and breathable. The feature extends the possibilities for micro-CHP applications. For experimental work, a simple convection flame plays on the plane cylinder end, the exhaust passing radially outwards through holes in an extension to the cylinder before being ducted away.

Simulation of the cycle of operations promises to be a straightforward matter of modifying an existing computer model of a Ringbom engine. At the time of writing this remains to be done, but interested readers may care to look out for an account in the Proceedings of the International Stirling Engine Conference, Osnabruek, September 2001 to be published jointly by ECOS GmbH and the Verein Deutscher Ingenieure.

14

In praise of Robert Stirling

14.1 Citation

Stirling's engine preceded that of Sadi Carnot by 8 years. It relied on the principle of reversible heat exchange, was patented, built and put into service. In a more genteel era (1884), Fleeming Jenkin put it this way:

'You have, therefore, a theoretically perfect heat engine, every portion of the action of which is reversible, and reversibility, as you have no doubt been told in previous lectures, is the test of a perfect engine. Stirling's is a perfect engine and it is the first perfect engine ever described.'

(In fact, the inevitable element of thermal conduction is *irreversible*, but Jenkin's meaning is clear.)

Carnot's engine, by contrast, is an abstraction, was not patented, was not built and never will be. The Carnot cycle is celebrated and known to every student of engineering and physics. Where the Stirling cycle gets a mention it is misunderstood and misrepresented, even down to the fourth and latest edition of the UK's most widely used teaching text on engineering thermodynamics (Rogers and Mayhew 1992).

This is an historical wrong. However, if the air engine re-emerges in an important commercial role such as CHP there will be hope for a long-overdue reappraisal of Stirling's unique scientific contribution. It is a tragedy that the written records are not available. Had they been published there might have been fewer Stirling engines and regenerators designed by the seat of the pants, and correspondingly fewer expensive disappointments. The paucity (Edelman 1969) of such biographical material is likely to delay conventional recognition of this uniquely gifted man.

On this basis alone the search for documentary material is urgent: such matters can frequently be speedier for being focused, and a focus can sometimes be found where an anomaly or injustice has shown up. Aiming for a wider recognition for the inventor might have a self-fulfilling effect. A man who produces a brilliantly crafted patent which is at the same time a document of monumental scientific importance *is a communicator and a writer*. When the rediscovery of the London patent was celebrated in 1917, the Librarian at the Patent Office, a Mr Hulme, wrote (1917) that *'had Stirling been a patent agent instead of a divine and a scientist he would have had equal eminence'*.

The records may well be languishing in an attic somewhere, just as the London patent had been. The larger the number of people who have heard the name, the greater will be the chance of material coming to light, and of due standing for Robert Stirling.

Or will it? This book opened its treatment of the regenerator (Chapter 1, Section 1.3) with a quotation from Professor Fleeming Jenkin's celebrated lecture (*op. cit.*), which thus concludes its discussion of the Stirling engine: *'You have here one and the same man devising an engine which is theoretically perfect, and devising the regenerator.'* What heightened standing could come of further forensic enquiry? More to the point would be that research and development in the twenty-first century should be worthy of the original concept.

14.2 How might the unique genius of Robert Stirling be celebrated?

The sophisticated, silent, highly efficient Stirling engines currently under development for CHP may eventually become Stirling's lasting memorial. For this to be so, it will be necessary to ensure that they are called after the inventor rather than being hijacked to some proprietary trade name or contemporary jargon.

It is to be hoped that academics will face their responsibility of enquiring into the functioning of the regenerator and the Stirling engine when editing future editions of their textbooks.

An appropriate celebration of Stirling's legacy would be to link the new Millennium, the new Scottish identity and the 200th anniversary of the patent in 2016 through a full-size working model. At 10 ft high it would have the scale of a statue on a plinth, while offering the added interest of being 'live'. A thermodynamic and kinematic simulation is already in existence, and a set of engineering drawings could follow rapidly and cheaply from the Pro-Engineer model. The cost might be little more than that of a passive bronze statue. The engine's operational days could be revenue earning. Some might consider such a memorial of greater social, cultural, and scientific substance than the Millennium Dome.

14.3 A task completed – or barely begun

Although addressing a different era in each case, successive chapters of this book have asked the same question: has the thermodynamic personality of the air engine been probed in sufficient depth to allow design and quantity production of a commercially competitive engine? As the Stirling engine nears the bicentenary of the original conception the question becomes especially pressing.

If so, what challenges remain? Cost reduction will no doubt be a priority, so the regenerator will remain a focus of attention for some time to come. There will be pressure to find alternatives to the inherently expensive wire-mesh type, and to extract the ultimate in performance where wire mesh is the sole option. Heat transfer and flow friction correlations will be required for the alternative matrix materials. The Kays and London correlations which have served so well for half a century will need to be re-examined: these were acquired (Section 5.8) using temperature-measuring equipment which is nowadays to be found only in museums, i.e. manually adjusted potentiometer and mirror galvanometer. Use of the Schumann solutions to invert the results embodies the assumption that the fluid temperature signal achieved in the laboratory is the step input of the original analysis. Moreover, the slow time–response of the instrumentation imposed the need to locate and match steady gradients between experimental and analytical output signals. These aspects are critical: if the experimental temperature wave is not a true step, then the inversion process imposes an error of uncertain magnitude (Kohlmayr 1966).

The original experiments may now be substantially improved upon, the uncertainties all but eliminated, and the whole process automated. Electronic data acquisition in real time means that the output temperature signal may be captured accurately. The availability of a general solution to regenerator transient response means that the output signal may be interpreted in terms of *any* input waveform: the difficult laboratory task of achieving a step input is no longer necessary.

Even without these necessary developments, the evidence is that sufficient has at last been achieved to give retrospective justification to earlier claims (deBrey *et al.* 1947) for the role of *modern knowledge about heat transfer, flow resistance, etc*. The Viebach approach is based squarely on these accumulated thermodynamic insights, and the results speak for themselves: high efficiency and low exhaust pollution levels in a concept well-balanced in relation to the proposed application. If cost expectations are realized, reliability and the rest may be expected to follow in the time-honoured fashion – by evolutionary development. Indeed, it is a realistic hope that the year 2016, just two centuries on, will be a time for a celebration of this remarkable technology in all its aspects – and above all in its widespread domestic use: air-engine-powered CHP.

Appendix

Literary output of Theodor Finkelstein

The structure of this book, with the contribution of the senior author focused on the classic treatment of air engines, gives a distorted picture of his contribution to the modern science of Stirling engine analysis and design. To redress the balance, this Appendix documents his literary output.

Finkelstein, T. (1952). Self-acting cooling cycles. Thesis submitted for the Diploma of Imperial College, Imperial College of Science and Technology, London.

- (1952). Theory of air cycles with particular reference to the Stirling engine. PhD dissertation, London University.

- and **Polonski** (1959). Development and testing of a Stirling cycle machine with characteristics suitable for domestic refrigeration. Report No. W/M(3A) u.5, The English Electric Company, Mechanical Engineering Laboratory, Whetstone.

- (1959). Air Engines. *The Engineer*, London (March 27, 492–497; April 3, 522–527; April 10, 568–571; May 8, 720–723).

- (1961). Generalized thermodynamic analysis of Stirling engines. Trans. Soc. Automotive Engineers Winter Annual Meeting, January 11–15, Detroit, Michigan, paper 118B.

- (1961). Optimization of phase angle and volume ratio for Stirling engines. Trans. Soc. Automotive Engineers Winter Annual Meeting, January 11–15, Detroit, Michigan, paper 118C.

- (1961). Regenerative thermal machines. *Battelle Technical Review*, Columbus, Ohio, May.

- (1961). Cyclic processes in closed regenerative gas machines analyzed by a digital computer simulating a differential analyzer. *Trans. ASME*, **84**, Series B, p. 165–179, June 14. Reprinted in *Trans. ASME*, **84**, Series B, No.1, February 1962; also in: *J. of Engng for Ind.*, 165, February 1962.

- (1961). Internally-focusing solar power systems – conversion of solar energy into power. Proc. ASME Winter Annual Meeting, New York, NY, November 26–December 1.

- (1962). Regenerative thermal machines. *International Science and Technology*, New York, May.
- (1962). Simulation of a regenerative reciprocating machine without mechanical wear on an analog computer. *Atomics International Technical Data Record* TDR-AI-8732, July 15.
- (1963). Preliminary design of a 500 W linear electromagnetic generator of the doubly-excited compensated type. *Atomics International Technical Data Record* AI-64-TDR-9323, December 16.
- (1963). Thermoelectrodynamic Oscillating Piston Power Conversion System ('TOPPS'), *Atomics International Technical Data Record* AI-64-TDR-21, January 30.
- (1964). Experimental prototype of heat-actuated vibrator. *Atomics International Technical Data Record* AI-64-TDR-21, January 30.
- (1964). Analysis of practical reversible thermodynamic cycles. Paper 64-HT-37 Joint American Inst. of Chemical Engineers and ASME Heat Transfer Conf., Cleveland, Ohio, August 9–12.
- (1964). Analog simulation of Stirling engines. Proc. Western Simulation Council Meeting, Los Angeles, California, September 1963; also in *Simulation*, No. 2, March.
- (1964). Specific performance of Stirling engines. 3rd Conf on Performance of High Temperature Systems, Pasadena, California, December 7–9. Also in book: Gordon and Breach, Science Publishers, New York.
- (1965). Simulation of a regenerative reciprocating machine on an analog computer. Paper 949F, SAE Int., Engineering Congress and Exposition, Detroit, Michigan, January 11–15.
- (1967). Thermophysics of regenerative energy conversion. Paper 67-216, Am. Inst. Aeronautics and Astronautics, 5th Aerospace Sciences Meeting, New York, January 23–26.
- (1967). Thermodynamic analysis of Stirling engines. Paper 67-216, Proc. Am. Inst. Aeronautics and Astronautics, 5th Aerospace Sciences Meeting, New York, January 23–26. Also in *J. Spacecraft and Rockets*, **4**, (6), November.
- (1970). Thermocompressors, Vuilleumier and Solvay machines. Paper 709025, Proc. 5th Intersociety Energy Conversion Engineering Conference, Las Vegas, Nevada, September 21–25.

- (1973). Advanced theory of the Stirling cycle. Proc. Stirling Engine Technology Meeting, Van Karman Institute for Fluid Dynamics, Brussels, February 12–16.

- (1975). Computer analysis of Stirling engines. Proc. 10th Intersociety Energy Conversion Engineering Conf., Delaware, August 17–22.

- (1975). Computer analysis of Stirling engines. *Adv. in Cryogenic Engineering*, **20**, 269, Plenum Press, New York and London.

- (1977). Modelling, design and optimization of Stirling engines. Paper 779252, Proc. 14th Intersociety Energy Conversion Engineering Conf.

- (1978). Stirling cycle engines: definitions and thermodynamic fundamentals. *Proc. Inst. Gas Tech*, Rosemont, Illinois, June 14–15.

- (1978). Balanced compounding of Stirling engines. Paper 789194, Proc. 13th Intersociety Energy Conversion Engineering Conference, San Diego, California, August 20–25. Also as Paper SAE-P-75-CH1372-2-ENERGY in Proc. SAE.

- (1980). Analysis of a heat-activated Stirling heat pump. Paper 809363, Proc. 15th Intersociety Energy Conversion Engineering Conf., Seattle.

- (1984). Balanced compounding of Stirling engines – preliminary designs. Proc. 2nd Int. Conf. on Stirling Engines, Shanghai, China, June 21–24.

- (1992). Analysis of integrated heat pumps by the 'null engine' concept. Paper 929497, Proc. 27th Intersociety Energy Conversion Engineering Conf., San Diego, California, August 20–25, **5**, p. 155.

- (1992). Integrierte Vuilleumier Wärmepumpen mit Kräfteausgleich. (Integrated Vuilleumier Heat Pumps with Balanced Compounding). Trans. Europäisches Stirling Forum Osnabrück.

- (1992). Isothermal sinusoidal analysis of a balanced compounded Vuilleumier heat pump. Paper 929498, Proc. 27th Intersociety Energy Conversion Engineering Conf., San Diego, California.

- (1993). Optimization of volume ratios for Stirling cycle machines. Trans. Engineering/Economics Colloq., Obninsk, Russia, October 25–27.

Organ, A.J. and **Finkelstein, T.**

(1993). Stirling engine thermodynamic design – the missing link. Paper 93492 Intersociety Energy Conversion Engineering Conf., Boston, Massachusetts.

Finkelstein, T. (1994). Thermodynamische Betrachtungen: der Begriff der 'kumulativen Massen' und 'Thermodynamischen Mittelebene und die daraus abzuleitenden theoretischen Rauminhaltsbeziehungen (Thermodynamic considerations: the concepts of cumulative mass and thermodynamic mid-plane and the volumetric relationships derived therefrom). Proc. European Stirling Forum, Osnabrück, March 22–24.

- (1994). Insights into the thermodynamics of Stirling cycle machines. Paper AIAA-94-3951-CP, 29th Intersociety Energy Conversion Engineering Conf., Monterey, California.

- (1995). Gas particle trajectories in Stirling cycle machines. Proc. 7th Int. Stirling Engine Conf., Tokyo, November 2–6.

- (1995). (B.-H. Wen and N. Isshiki) Proposal for a simple Stirling engine with a multi-fuel agricultural drum incinerator. Proc. 7th Int. Stirling Engine Conf., Tokyo, November 2–6.

- (1996). Rechnungshilfe zur Erläuterung der Funktion und des Wirkungsgrads von Vuilleumiermaschinen. (Computational aid for the simulation of the operation and performance level of Vuilleumier machines.) Proc. 3rd European Stirling Engine Forum, Osnabrück, February 26–28.

- (1996). Optimization of the phase angle in ideal Stirling engines using the concept of 'tidal' and 'ancillary' domains. Proc. 31st Intersociety Energy Conversion Engineering Conf., Washington DC, August 11–16.

- (1997). Determination of Stirling cycle machine performance based on a new normalized system view. Proc. 8th Int. Stirling Conf., Ancona, Italy, May 27–30.

- (1998). Specific performance of various Stirling engine designs. Proc. 4th European Stirling Engine Forum, Osnabrück, February 24–26.

- (1998). A new isothermal theory for Stirling machine analysis and a volume optimization using the concept of 'tidal' and 'ancillary' domains. *Proc. Instn Mech. Engrs, Part C*, **212**, (May), 225–236.

- (1999). (With T. Ertle) Thermodynamics of an engine with an open regenerative cycle. Proc. 9th Int. Stirling Engine Conf. and Exhib., June 2–4, Pilanesburg, South Africa.

- (2000). Evaluation and simulation of the Stirling engine and of its performance. Proc. European Stirling Forum, Osnabrück, February 22–24.

References

Many references in the bibliography of the original *Air Engines* were somewhat cryptic. Where it has been possible to amplify, the fuller form is given here. Terminal numerals in italic square parentheses *[ ]* cross-reference to the original bibliography. (The process leads to more than one reference carrying the same numeral.)

Book titles are *italicized.*

Anon. (1834a). Ueber die neue von Herrn Ericsson erfindene Waermestoffmaschine, Caloric-Engine genannt. (From *Mechanics Magazine* No.535, 82). *Dinglers Polytechnisches Journal*, **51**, 81–84 *[25]*.

Anon. (1834b). Ueber Ericsson's erfundene Waermestoffmaschine. *Dinglers Polytechnisches Journal*, **51**, 173–185 *[25]*.

Anon. (1852a). Maschine zur Erzeugung von Triebkraft vermittelst der Ausdenhnung atmosphaerischer Luft durch die Waerme; patentiert fuer Edward Dunn in London am 26 Dezember 1850. (From *London Journal of Arts*, September 1851, 253) *Dinglers Polytechnisches Journal*, **123**, 86–90 *[30]*.

Anon. (1852b). Die Luftdruckmaschine des amerikanischen Capitaens Ericsson. *Dinglers Polytechnisches Journal*, **126**, 153–154 *[34]*.

Anon. (1853a). Die Maschineneinrichtung von Ericsson's Caloric-Schiff. *Dinglers Polytechnisches Journal*, **128**, 174–176 *[34]*.

Anon. (1853b). Ueber Ericsson's Caloric-Maschine. (From Norton W. R.: *Sillimans American Journal of Science and Arts*, May 1853, 393) *Dinglers Polytechnisches Journal*, **129**, 186–194 *[34]*.

Anon. (1855). Napier und Rankine's patientierte Luftexpansionsmaschine. (From *Mechanics Magazine*, 1854, No.1628.) *Dinglers Polytechnisches Journal*, **135**, 241–246 *[35]*.

Anon. (1856). Capitaen Ericsson's neue Luftexpansionsmaschine. (From *Mechanics Magazine*, 1856 No.1691). *Dinglers Polytechnisches Journal*, **140**, 259–265 *[32]*.

Anon. (1860a). Ueber die Verwendung der Ericsson'schen Maschinen in Schweden, und eine neue Verbesserung derselben. *Dinglers Polytechnisches Journal*, **158**, 394–395 *[32]*.

Anon. (1860b). Ericsson's neue calorische Maschine fuer kleine Kraftbeduerfnisse. (From *Wuerttembergischer Gewerbeblatt*, 1860, No.32) *Dinglers Polytechnisches Journal*, **157**, 162–166 *[32]*.

Anon. (1860c). Beschreibung der neuen calorischen Maschine von Ericsson. *Dinglers Polytechnisches Journal*, **157**, 321–322 *[32]*.

Anon. (1860d). American air engine. *Inst. Engrs in Scotland Trans.*, **4** 13–15.

Anon. (1860e). Ericsson's einfach wirkende calorische Maschine. *Zeit. des Ver. D. Ing.*, **V**, Heft 2, 3, 75–77 *[33]*.

Anon. (1862a). *The Engineer* (Description and illustration of Shaw's air engine) p. 138 (February 28); (Description of Wilcox' air engine) p. 288 (May 9) *[23]*.

Anon. (1867). Weitere Mitteilungen ueber die Heissluftmaschine von Lauberau in Paris. *Dinglers Polytechnisches Journal*, **185**, 423–426 *[50]*.

Anon. (1875). *Engineering*, **XIX**, 200–201 (March 12); 241–242 (March 26); 287–289 (April 9); 320 (April 16); 355–356 (April 30) *[21]*.

Anon. (1876). *Engineering*, **XXII**, 14 (July 9) *[56]*.

Anon. (1881). *Engineering*, **XXXI**, 459 (May 6) *[59]*.

Anon. (1883). *Engineering*, **XXXIV**, 368 (magnificent plate at p. 372) (April 20) *[18]*.

Anon. (1899). *Appleton Cyclopaedia of Applied Mechanics*, New York *[55]*.

Anon. (1917a). The Centenary of the Heat Regenerator and the Stirling Air Engine. *The Engineer*, December 14, 516–517 *[5]*.

Anon. (1917b). The Stirling Specifications. *The Engineer*, December 28, 567 *[42]*.

Anon. (1947). The Philips Air Engine. *The Engineer*, December 12, 549–550; and December 19, 572–579 *[61]*.

Anon. (1998). Drying in style. *Which?* Magazine, 30–33 (September 30).

Abbott, M.B. (1966). *An Introduction to the Method of Characteristics*. Thames and Hudson, London.

Adkins, C.N. and **Liebeck, R.H.** (1983). Design of optimum propellers. Paper AIAA-83-0190, AIAA 21st Aerospace Sciences Meeting, Reno, Nevada (January 10–13).

Alexander, J.M. and **Brewer, R.C.** (1963). *Manufacturing Properties of Materials*. van Nostrand, London.

'Artificer' (1940). Practical notes on hot-air engines. *The Model Engineer*, 206–208 (September 12); 235–237 (September 19); 263–264 (October 3); 280–281 (October 10); 295–297 (October 17); 310–312 (October 24); 336–337 (October 31); 363–366 (November 7); 375–376, 379 (November 14); 400–402 (November 21); 425–426 (November 28); 442–444 (December 5).

Bejan, A. (1982). *Entropy generation through Heat and Fluid Flow.* John Wiley, New York.

Beldam-Crossley (1991). Product data sheet 315 on Rulon LD 'plastic sealing material'. Beldam-Crossley Ltd, Bolton.

Beldam-Crossley (1995). Self-lubricated plane (*sic*) bearings – expanding the limits of design. Tech. pamphlet LM-SK-7-95, Beldam-Crossley Ltd, Bolton.

Blum (1861). Ueber die Wirkung der Luft in der calorischen Maschine. *Dinglers Polytechnisches Journal*, **159**, 404–406 *[32]*.

Boëtins, H. (1861). Die Ericsson'sche calorische Maschine. *Dinglers Polytechnisches Journal*, **159**, 82–89 *[32]*.

Boëtins, H. (1862). Die Ericsson'sche calorische Maschine und Leucir's Gasmaschine, *Zeit. des Vereins D. Ing.*, **IV**, Heft 1, 66–69 *[32]*.

Bopp, G. (and Co.) (undated). Woven wire cloth. Bopp documentation G/0100/02, G. Bopp and Co. AG, Bachmannweg 20, 8046 Zurich, Switzerland. UK representative G. Bopp and Co. Ltd, London, UK.

Bourne, J. (1878). *Examples of Steam, Air and Gas Engines of the Most Recent Approved Type*. Longmans Green and Co., London *[12]*.

Bradley, A. (1974). Demonstration of engine and personal exchanges with author, who now holds dossier of drawings, photographs and correspondence on the development of the engine donated by Prof. G. Walker of Calgary University.

Bratt, C. (2000). Development and production of Stirling engines for submarine propulsion. Proc. European Stirling Forum, Osnabrueck, pp. 313–318.

Briot, C. (1869). *Theorie Mecanique de la Chaleur*, pp. 78–90, Gauthier-Villars, Paris.

Brooman, R.A. (Ed) (1854a). Shaw's American hot-air engine. *The Mechanics' Magazine*, **LXI**, No.1616 (July 29), 97–99.

Brooman, R.A. (Ed) (1854b). The 'Ericsson' transformed to a steamer. *The Mechanics' Magazine*, **LXI**, No.1619, 179.

Brooman, R.A. (Ed) (1854c). Napier and Rankine's patent hot-air engine. *The Mechanics' Magazine*, **LXI**, No.1628 (October 21), 385–389 *[27]*.

Brown, L. (Ed) (1993). *Shorter Oxford Dictionary*, Oxford.

Carlsen, H. (2000). Design and performance of a new 9kW Stirling engine. Proc. European Stirling Forum, Osnabruck, pp. 183–191.

Carnot, S.N.L. (1824). *Reflexions sur la Puissance Motrice de Feu et sur les machines propres a developper cette puissance*. Bachelier. English translation by R.H. Thurston, *Am. Soc. Mech. Engrs*, (1943) *[4]*.

Cayley, Sir G. (1807). Description of an engine for affording mechanical power from air expanded by heat. Article IV (*Nicholson's*) *J. Nat. Phil. and the Arts*, **XVIII**, 260–262 (December) *[14]*.

Cheverton, B. (1852). On the use of heated air as motive power. *Proc. Inst.Civ. Engrs*, **XII**, 312–351 *[6]*.

Church, W.C. (1890). *The Life of John Ericsson,* vols I and II. Sampson Low Marston Searle and Riviginton Ltd, Fleet St, London *[8]*.

Clapham, E. (1977). *Model Engineer*, 852–856, 908–912.

Clapham, E. (1999). Presentation to UK Stirling Engine Society, Salford Priors, October.

Clucas, D.M. (2000). Personal communication with author (e-mail), 28 August.

Clucas, D.M. and **Raine, J.K.** (1994). Development of a hermetically-sealed Stirling engine battery charger. *Proc. Inst. Mech. Engrs*, **208**, 357–366.

Codegone, C. (1955). Sur le cycle Frigorifique Philips. Comtes Rendus du IX Congrès Internat. du Froid. Paris, August 31 – September 7, pp. 3025–3031 *[70]*.

Combes. (1853). Bericht ueber die Warmluft-Maschine des Capitaens Ericsson. (From Combes: Annales des Mines, Fuenfte Reihe 1853, Band **III**, 775) Band **130**, 81 *[34]*.

Coppage, J.E. (1952). PhD dissertation, Stanford University. Published as Technical Report No.16 Office of Naval Research. Stanford University., California (December 1).

Coppage, J.E. and **London, A.L.** (1956). Heat transfer and flow friction characteristics of porous media.. *Chem. Engng. Progress*, **52** (2), 57–63.

Creswick, F.A. (1965). Thermal design of Stirling cycle machines. Paper 949C, Proc. Int. Automotive Engng. Congr., SAE, Detroit, January 11–15.

Cuttil, W.E., **Malik, M.J.**, and **Decker, V.L.** (1962). Investigation of a 3kW Stirling cycle solar power system. Vol. IX, Power system control analysis. Technical Report WADD.TR-61-122 by Flight Accessories Lab., Aeronautical Systems Div., Air Force Systems Command, Wright Patterson Air Base, Ohio (March).

Dando, R. (1997). Personal communication with the author.

deBrey, H., **Rinia, H.**, and **van Weenen, F.L.** (1947). Fundamentals for the Development of the Philips Air Engine. *Philips Tech. Rev.*, **9**, No. 4, 97–104 *[63]*.

Delabar, G. (1864). Die Lauberau-Schwartzkopf'sche Heissluftmaschine. *Dinglers Polytechnisches Journal*, **172**, 81–87 *[46]*.

Delabar, G. (1865a). Belon's Heissluftmaschine und die damit von Tresca in Paris ausgestellten Versuche. *Dinglers Polytechnisches Journal*, **177**, 414–423 *[22]*, **194**, 257.

Delabar, G. (1865b). Die Heissluftmaschine von Windhausen und Huch. *Dinglers Polytechnisches Journal*, **178**, 169–175 *[22]*.

Delabar, G. (1865c). Roper's Heissluftmaschine. *Dinglers Polytechnisches Journal*, **178**, 249–251 *[22]*.

Delabar, G. (1866). Ueber die weiteren Verbesserungen an Heissluftmaschinen, *Dinglers Polytechnisches Journal*, **179**, 329–346 *[47]*.

Donkin, B. (1911). *A Text-book on Gas, Oil and Air Engines*, Fifth edition. Charles Griffin and Co., London *[20]*.

Drela, M. (undated). X-foil. Software for 'first-principles' design and display of airfoil sections.

Eckerth, W. (1869a). *Vierteljahresschrift des Deut. Ing. und Arch. Ver.*, **1**, Jahrgang, Heft 2 *[51]*.

Eckerth, W. (1869b). *Technische Blätter*, 1 Jahrgang, Prague *[52]*.

Edelman, H.C.M. (1969). In search of Stirling. *Announcer*, **23**, No. 1, 6–12.

Ericsson, J. (1855). Ueber die calorische Maschine (From *Mechanics Magazine*, July 1855, 1665), *Dinglers Polytechnisches Journal*, **159**, 315–316 *[35]*.

Ericsson, J. (1876). *Contributions to the Centennial Exhibition*. Printed for the author by the Nation Press, New York *[54]*.

Ewbank, T. (1854). Thoughts on the Caloric engine. *The Mechanics' Magazine*, **LXI**, 1629, 411–414.

Ferguson, E.S. (1961). John Ericsson and the age of caloric. Paper 20 from Contributions from the Museum of History and Technology, pp. 41–60. Bulletin 228, US National Museum, Smithsonian Institution, Washington DC.

Feulner, P. (1999). Personal communciation with the author.

Feurer, B. (1973). Degrees of freedom in the layout of Stirling engines. Proc. Stirling Engine Short Course, von Karman Inst. Fluid Dynamics, Brussels, February 12–16.

Finkelstein, T. (1959). Air engines. *The Engineer*, (27 March) 492–497; (3 April) 522–527; (10 April) 568–571; (8 May) 729–733.

Finkelstein, T. (1960a). Optimization of the phase angle and volume ratio of Stirling engines. Paper 118C, SAE Winter Annual Meeting.

Finkelstein, T. (1960b). Generalized thermodynamic analysis of Stirling engines. Paper 118B, SAE Winter Annual Meeting.

Finkelstein, T. (1961). Regenerative thermal machines. *Battelle Tech. Rev.* (pp. 1–7 in reprint form).

Finkelstein, T. and **Eibling, J.A.** (1961). Internally-focusing solar power systems. Paper 61-WA -297, Proc. ASME Winter Annual Meeting, New York, 26 November – 1 December.

Finkelstein, T. (1963a). Preliminary design of a 500W linear electromagnetic generator of the doubly-excited compensated type. Atomics International Technical Data Record AI-64-TDR-9323, December 16.

Finkelstein, T. (1963b). Thermoelectrodynamic Oscillating Piston Power Conversion System ('TOPPS'), Atomics International Technical Data Record AI-64-TDR-21, January 30.

Finkelstein, T. (1964). Experimental prototype of heat-actuated vibrator. Atomics International Technical Data Record AI-64-TDR-21, January 30.

Fokker, H. and **van Eekelen** (1978). Typical phenomena of the Stirling cycle as encountered in a numerical approach. Proc. 13th Inter-Society Energy Conv. Engng Conf., SAE, Detroit, Michigan, USA, pp. 1746–1752.

Fox, R. (1986). *Sadi Carnot – Reflexions on the Motive Power of Fire.* Manchester University Press, Manchester.

Galloway, E. (1830). *History and Progress of the Steam Engine (with a Practical Investigation of the Structure and Applications)*, p. 667, (Printed for) Thomas Kelly, London *[44]*.

Gibbs-Smith, C. (1957). The Father of the Modern Aeroplanes. *The Listener*, December 19 *[15]*.

Grashof, F. (1860). Theorie der Ericsson'schen calorischen Maschine. *Zeit. des Ver. D. Ing.*, **IV**, Heft 12, 292–301 *[33]*.

Grashof, F. (1890). *Theoretische Maschinenlehre,* Band III: *Theorie der Kraft-maschinen*, Hamburg *[2]*.

Greenwood, A. (2000). Personal telephone communication with the author 23 August.

Hamaguchi, K. and **Yamashita, I.** (1995). Design approach of Stirling engine using scaling rules. Paper ICSC-95 038, Proc. 7th. Int. Conf. on Stirling cycle machines. Japan. Soc. Mech. Engrs, Tokyo, pp. 523–528.

Hargreaves, C. M. (1991). *The Philips Stirling Engine*, Elsevier Science, Amsterdam.

Hausen, H. (1927). Habilitationsschrift (qualifying dissertation). Tech. Hochsch. Muenchen.

Hausen, H. (1929). Über die Theorie des Wärmeaustausches in Regeneratoren (On the theory of heat exchange in regenerators). *Z. angewandte Math. und Mech*, **9**, June 3, 173–200.

Hausen, H. (1930). *Tech. Mech. Thermodynam*, **1**, 219.

Hausen, H. (1950). *Wärmeübertragung in Gegenstrom, Gleichstrom und Kreuzstrom. (Heat exchange in counter- parallel- and cross-flow)*, Springer, Berlin.

Hirsch, M.J. (1873). Theorie des machines aerothermiques. *Annales des Ponts et Chaussees*, **14**, 409–528 (with plate No.7).

Hulme (1917). The Regenerator, *The Engineer*, December 14, 523 *[40]*.

Jakob, M. (1957). *Heat Transfer*, Vol. II, Wiley, New York.

James, R.G. (1999). Personal communications with the author.

Jenkin, H.C. Fleeming (1885). Gas and caloric engines. Lectures delivered to Inst. Civ. Engrs, on 21 February 1884. Published in *Heat and its Mechanical Applications*, Inst. Civ. Engrs, London *[13]*.

Kalinin, E.K. and **Dreitser, G.A.** (1994). Unsteady convective heat transfer in channels. *Adv. Heat Trans.* **25**, 1–150.

Kays, W.M. and **London, A.L.** (1964). *Compact Heat Exchangers*, McGraw-Hill, New York.

Kennedy, R. (1904–5). *Modern Engines and Power Generators*, Vol. II (of 6), Caxton, London, *[17]*.

Kirk, A.G. (1873). Mechanical Production of Cold. *Proc. Inst. Civ. Engrs*, **37**, 244 *[65]*.

Kiyoshi, K. (1990). Warm air generating apparatus. Patent application number JP 2283307, Toshiba Corp as applicant (20 November).

Knoke, J.O. (1899). *Die Kraftmaschine des Kleingewerbes*, Berlin *[11]*.

Köhler, J.W.L. and **Jonkers, C.O.** (1954a). Fundamentals of the gas reriger-ating machine. *Philips Tech. Rev.*, **16**, No. 3, September, 69–78 *[66]*.

Köhler, J.W.L. and **Jonkers, C.O.** (1954b). Construction of a gas refrigerat-ing machine. *Philips Tech. Rev.*, **16**, No. 4, October, 115–116 *[67]*.

Köhler, J.W.L. and **Jonkers, C.O.** (1954c). De Kindgaskoelmachine. *De Ingenieur*, **49** December 3 *[68]*.

Köhler, J.W.L. and **van der Steer, J.** (1955). A small liquid nitrogen plant using gas refrigerating machine. Comtes Rendus du IX Congrès Internat. du Froid, Paris August 31–September 7, pp. 1057–1063 *[69]*.

Köhler, J.W.L. (1955). Principles of gas refrigerating machines. *Nature*, October 1, 623–626 *[71]*.

Kohlmayr, G.F. (1966). Exact maximum slopes for transient matrix heat trans-fer testing *Int. J. Heat and Mass Trans.*, **9**, 671–680.

Kolin, I. (1991). *Stirling Motor*. University Press, Zagreb.

Kooke, C. (1873). On Wenham's heated air engine. *Proc. Instn Mech. Engrs*, 63–85, London *[24]*.

Lanchester, F.W. (1898). Improvements in fluid pressure engines. UK Patent No.10836.

Larque, I. (1995). Design of a small 30 cc pressurized Ross-drive Stirling engine. Paper ICSC-95 082, Proc. 7th. Int. Conf. on Stirling cycle machines. Japan Soc. Mech. Engrs, Tokyo, pp. 523–528.

Larrabee, E. (1979). Design of propellers for motor-soarers. Un-annotated report – possibly a version of 'Practical design of minimum induced loss propellors' SAE business aircraft meeting and exposition, Wichita, Kansas, April.

Lawrie, J.G. (1860). On the operation of air engines. *Inst. Engrs. in Scotland Trans.*, **4**, 141–158.

Lo, C.-B. (1999). Appraisal of a 25 cc Stirling cycle engine. MPhil dissertation, University of Cambridge.

Lundholm, S. G. K. (1983). The design and development of a 40kW Stirling engine – the power system. Paper C27/82, Proc. Inst. Mech. Engrs. Conf. Stirling engines – Progress Towards Reality. University of Reading, 25–26 March.

Manby, E. (1853). On the Caloric Engine. *Proc. Inst. Civ. Engrs*, **XII**, 558–563 *[43]*.

Martini, W. (1978). Design manual for Stirling engines. DOE/NASA report, available through US National Technical Information Service (NTIS).

Matschoss, K. (1925). *Männer der Technik*. Berlin *[38]*. Publisher unknown, but see also (1937) *Grosse Ingenieure – Lebensbeschreibung aus der Geschichte der Technik*, Muenchen/Berlin, pp. 186–196, and/or the translation by H. S. Hatfield (1939), *Great Engineers,* G. Bell and Sons, pp. 202–216.

Meek, J. (2000). Home-made answer to generating electricity harks back to the past. *The Guardian*, 2 September.

Meijer, R.J. (1960). The Philips Stirling thermal engine. PhD dissertation, Tech. Univ. Delft.

Meijer, R.J. (1965). Philips Stirling engine activities. Paper 949E, Proc. Int. Automotive Engng. Congr., SAE, Detroit, Michigan.

Monson, D.S. (1961). Second quarterly Tech. Progress Rep. Continuation of 3 kW Stirling solar power system program. Report under contract AF33(616)-8332 Project No.3145, Task No.30500 by Allison Div. of General Motors Corp., Indianapolis, Indiana. (D.S. Monson named as project engineer, i.e. not necessarily author).

mRT (1993). *Scalit* gas circuit design software. Regenerative Thermal Machines, Cambridge CB3 8BH.

Miyabe, H. *et al.* (1982). An approach to the design of Stirling engine regenerator matrix using packs of wire gauzes. Paper 829306, Proc. 17th Inter-Societies Energy Conversion Engineering Conf. (IECEC), pp. 1839–1844.

Norton, W.R. (1853). Ueber Ericsson's Caloric-Maschine. (From Silliman's *American Journal of Science and Arts*, May, 393) *Dinglers Polytechnisches Journal*, **129**, 186–194 *[34]*.

Nusselt, W. (1927). Die theorie des Winderhitzers. *Zeit. des Vereins deutscher Ing.*, **71**, 85.

Oon, A. (1998). Aesthetics and styling. Final year project report, Engineering Dept, University of Cambridge.

Organ, A.J. (1962). An appraisal of the hot-air engine. Final year project report, Dept Mech. Engng, University of Birmingham.

Organ, A.J. (1982). Gas dynamics of the temperature-determined Stirling cycle. *J. Mech. Engng Sci.*, **23**, (4) 207–216.

Organ, A.J. (1990). Mechanical design data book. Internal publication, Engineering Dept, University of Cambridge.

Organ, A.J. (1992a). *Thermodynamics and Gas Dynamics of the Stirling Cycle Machine*. Cambridge University Press.

Organ, A.J. (1992b). 'Natural' coordinates for analysis of the practical Stirling cycle. *Proc. Instn Mech. Engrs, Part. C*, **206**, 407–415.

Organ, A.J. (1993a). Steady-flow $C_f - N_{re}$ correlations for the wire gauze regenerator inferred from linear wave analysis. *Proc. Instn Mech. Engrs, Part C*, **207**, 53–62.

Organ, A.J. (1993b). Flow in the Stirling regenerator characterized in terms of complex admittance. I. Theoretical development. *Proc. Inst. Mech Engrs, Part C*. **207**, 117–125.

Organ, A.J. (1994a). Solution of the classic thermal regenerator problem. *Proc. Instn Mech. Engrs, Part C*, **208**, 187–197.

Organ, A.J. (1994b). The wire mesh regenerator of the Stirling cycle machine. *Int. J. Heat Mass Trans.*, **37**, 2525–2534.

Organ, A.J. (1997a). *The Regenerator and the Stirling Engine*. Mechanical Engineering Publications (MEP), London.

Organ, A.J. (1997b). Solution of the conjugate heat transfer problem. *Proc. Instn Mech. Engrs, Part C*, **211**, 17–24.

Organ, A.J. *et al.* (1998). Gas powered heat gun. Patent application lodged with UK Patent Office (12 April).

Organ, A.J. (2000a). 200 years of the thermal regenerator. *Proc. Instn Mech. Engrs. Part C*, **214** (Millennium Special Issue), 269–288.

Organ, A.J. (2000b). Stirling's air engine – a thermodynamic appreciation. *Proc. Instn Mech. Engrs*, **234** (C4), 511–536.

Organ, A.J. (2000c). A gas-powered, portable hair drier – an account of the design study. (Under review.)

Organ, A.J. and **Lo, C.B.** (2000). Thermodynamic personality of a 25 cc Stirling engine. Part 1 Design considerations, test installation and energy balance, pp. 39–47, Part 2 Hydrodynamic pumping loss and thermal short, pp. 49–57. Proc. European Stirling Forum, Osnabruck, February.

Organ, A.J. (2000d). Regenerator analysis simplified. Proc. European Stirling Forum, Osnabruck, February, pp. 27–37.

Organ, A.J. (2000e). Understanding the Stirling regenerator. Under review, *Proc. Instn Mech. Engrs, Part C*.

Organ, A.J. (2000f). In praise of Robert Stirling. Lecture delivered to the Stirling Seminar, Kew Bridge Steam Museum, 20 July. Under consideration for publication by the Newcomen Society. Pre-prints available by contacting the author allan_j_organ@talk21.com

Organ, A.J. (2001). The optimum screw propeller – extension to the static thrust case. Under review, *Proc. Instn Mech. Engrs, Part C*.

Organ, A.J. and **Rix, D.H.** (1993). Flow in the Stirling regenerator characterized in terms of complex admittance. II. Experimental investigation. *Proc. Instn Mech. Engrs, Part C*, **207** 127–139.

Orlebar, C. (1986). *The Concorde story*, Temple Press, London.

Patent Office (1995). Patent protection. Publication DTI/Pub 1507/30K/10.95/AR (October).

Percival, W.H. *et al.* (1960). GMR Stirling thermal engine part of the Stirling engine story – 1960 chapter. Proc. SAE Winter Annual Meeting, Detroit, January 11–15, published in *Trans. SAE*, **68**. 1964 re-print with discussions.

Phillips, R. (1833). *Dictionary of the Arts of Life and Civilisation*, London. Publisher not declared, but book dedicated to George Birkbeck MD, president of the London Mechanics Institution, so possibly published by them *[26]*.

Poingdestre, W.W. (1849). Description of Sir George Cayley's Air Engine. Paper No.822, *Proc. Inst. Civ. Engrs*, **IX**, 194–196 (with plate 9 and discussion) *[16]*.

Poppe, A. (1853). Ericsson's Luftexpansionsmaschine (caloric engine) und das ihr zu Grunde liegende *Prinzip, Dinglers Polytechnisches Journal*, **127**, 401–415 *[30]*.

Popplewell, W.C. (1897). *Heat and Heat Engines*, Manchester *[45]*.

Press, W.H. *et al.* (1989). *Numerical Recipes – the Art of Scientific Computing* (FORTRAN ed.), Cambridge University Press, Cambridge.

Prosses, R.B. (1917). The Centenary of the Heat Regenerator and the Stirling Air Engine. *The Engineer*, December 21, 537 *[41]*.

Rankine, W.J.M. (1854a). On the Means of Realising the Advantages of the Air Engine. Read to the Brit. Assoc. for the Adv. of Sc. at Liverpool (September 1884). Published in *Edinburgh New Philosophical J*, **1**, 1855 *[9]*.

Rankine, W.J.M. (1854b). On the geometrical representation of the expansive action of heat, and the theory of thermo-dynamic engines. Contribution No.VI, *Phil. Trans. Roy. Soc. Lond.* **144**, (1), 115–175, and in particular Section III – Of the efficiency of thermo-dynamic engines worked by the expansion and condensation of permanent gases, 137–146 and Section IV – Of the temperature . . . and the numerical computation of the efficiency of air engines, 146–154.

Rankine, W.J.M. (1908). *A Manual of the Steam Engine and other Prime Movers*. Charles Griffin and Co. Ltd, London *[10]*.

Rietlinger, E. (1873). Ueber Kreisprozesse mit zwei isothermischen Kurven. *Zeit. des Oesterreichischen Ingenieure und Architekten-Vereins*.

Rinia, H. (1946). New Possibilities for the Air Engine. *Proc., Koninklijke Nederlandsche Akad. van Wetenschappen*, February, pp.150–155 (published in English) *[1]*.

Rinia, H. and **duPré, F.K.** (1946). Air Engines. *Philips Technical Review*, **8** (5), (May), 129–136 *[62]*.

Rix, D.H. (1984). An enquiry into gas process asymmetry in the Stirling cycle machine. PhD dissertation, Engineering Department, University of Cambridge.

Robinson, E. (1902). *Gas and Petroleum Engines*, E. and F. N. Spon Ltd, London *[29]*.

Rogers, G.F.C. and **Mayhew, Y.R.** (1992). *Engineering Thermodynamics, Work and Heat Transfer*, fourth edition, Longman, London.

Röntgen, R. (1888). *The Principles of Thermodynamics with Special Application to Hot Air, Gas and Steam Engines*, (Trans. du Bois), New York *[31]*.

Ross, A. (1993). *Making Stirling Engines*. Ross Experimental, Columbus, USA.

Rühlich, I. (2000). Optimization of fluid flow and heat transfer in regenerators for Stirling machinery. Proc. European Stirling Forum, Osnabrück, February, pp. 149–154.

Schalkwijk, W.F. (1958). A simplified regenerator theory. *Trans. ASME, J. Engng Power*, 1–9, Paper 58-A-135.

Schalkwijk, W.F. (1960). A simplified regenerator theory. *Brit. Chem. Engng.* 33–34.

Schmidt, E.H. (1861). Versuche ueber Arbeitsstaerke und Brestoffsverbrauch der calorischen Maschine. *Dinglers Polytechnisches Journal*, **159**, 407–412 *[36]*.

Schmidt, G. (1861a). Theorie der geschlossenen calorischen Maschine von Laubroy und Schwartzkopff in Berlin. *Zeit. des Oester. Ing. Ver.*, 79 *[48]*.

Schmidt, G. (1861b). Theorie der geschlossenen calorischen Maschine von Laubroy und Schwartzkopff in Berlin. *Dinglers Polytechnisches Journal*, **160**, 401–416 *[48]*.

Schmidt, G. (1871). Theorie der Lehmann'schen calorischen Maschine. *Zeit. des Vereins deutscher Ing.*, **XV** (2), 1–12 (January), 98–112 (February).

Shaw, H.S.H. (1880). Small Motive Power. *Proc. Inst. Civ. Engrs*, **LXII**, 290 *[39]*.

Schöttler, R. (1881). Ueber die Heissluftmaschine von Rider. *Zeit. des Ver. Deutsch. Ing.*, **25**, Heft 11, 634–647 (November) *[57]*.

Schröter, M. (1883). Ueber die Anwendung von Regeneratoren bei Hessluft maschinen. *Zeit. des Ver. Deutsch. Ing.*, **27**, Heft 7, 450–459 (July).

Schumann, T.E.W. (1929). Heat transfer: a liquid flowing through a porous prism. *J. Franklin Inst.*, **208**, 1243–1248 (104th year), September, 405–416.

Shapiro, A.H. (1954). *Dynamics and Thermodynamics of Compressible Fluid Flow*, Vol. II, Ronald, New York.

Senft, J.R. (1993). *An Introduction to Stirling Engines*. Moriya Press, USA.

Slaby, A. (1878). Beiträge zur Theorie der Geschlossenen Luftmaschinen, *Verhandlungen des Vereins zur Befoerderung des Gewerb(e)fleisses* (Leonard Simion Verlag, Berlin) 57 Jahrgang, pp 375–409 *[3]*.

Spalding, D.B. and **Cole, E.H.** (1966). *Engineering Thermodynamics*. Second edition. Edward Arnold, London.

Stirling, J. (1845). Description of Stirling's improved air engine. *Proc. Inst. Civ. Engrs*, **IV**, 348–361 *[46]*.

Stirling, J. (1853). Contribution (at pp. 599–600) to discussion 'Heated Air Engine'. *Proc. Inst. Civ. Engrs*, **XII**, 558–600.

Stirling, P. (1860). Notes on Stirling's air engine. *Inst. Engrs in Scotland Trans.*, **4**, 40–50.

Stirling, R. (1816). Patent no. 4081 – Improvements for diminishing the consumption of fuel, and in particular an engine capable of being applied to the moving (of) machinery on a principle entirely new.

Takase, C.N. (1972). Otimisação do desempenho específico de motores Stirling. (Optimization of the specific performance of Stirling engines.) MSc dissertation, Inst. Tecnológico de Aeronáutica (ITA) CTA, São José dos Campos SP, Brazil.

Tew, R., **Jeffries, K.**, and **Miao, D.A.** (1978). A Stirling engine computer model for performance calculations. US DoE/NASA Report TM-78884 (July).

Tresca, H. (1867). Bericht ueber die neuesten Versuche mit einer Belon'schen Heissluftmaschine. *Dinglers Polytechnisches Journal*, **185**, Heft 6, 409–423 *[19]*.

Urieli, I. (1977). A computer simulation of Stirling cycle machines. PhD. dissertation, University of the Witwatersrand.

Urieli, I. and **Berchowitz, D.M.** (1984). *Stirling Cycle Engine Analysis*. Adam Hilger, Bristol.

van Weenen, F.L. (1947). The Construction of the Philips Air Engine. *Philips Tech. Rev.*, **9** (5), 125–134 *[64]*.

Walker, G. (1961). Operations cycle of the Stirling engine with particular reference to the function of the regenerator. *J. Mech. Engng Sci.*, **3** (4), 394–408.

Walker, G. (1980). *Stirling Engines*, Oxford.

Walker, G. (1981). The Stirling engine for railway locomotive propulsion. Report TP 2930E prepared under contract 4270 for Transport Canada, Transportation and Development Centre, Montreal.

Walker, G. (1999). Personal communication with the author (23 May).

Weisbach, J. *Mechanics of Engineering*. Trans. duBois *[49]*. There is a *Principles of the Mechanics of Machinery and Engineering*, Vols. I and II (1947) translation by L. Gordon and published in 2 vols by Hippolyte Bailliere, London/Paris, but Vol. II does not extend to the Art. 475 of the text *[49]*.

West, C.D. (1986). *Principles and Applications of Stirling Engines*, van Nostrand Reinhold, New York.

White, F.M. (1979). *Fluid Mechanics*, McGraw-Hill, Kogakusha, Tokyo.

Yamamoto, H., **Takahashi, S.**, **Ebine, S.** (1994a). Performance characteristics of Stirling engines – an expression based on law of Similarity. Part 1. Unidentified copy supplied by authors (in Japanese).

Yamamoto, H., **Takahashi, S.**, **Ebine, S.**, and **Enokida, M.** (1994b). Performance characteristics of Stirling engines – an expression based on law of Similarity. Part 2. Unidentified copy supplied by authors (in Japanese).

Yellott, J.I. (1957). Power from Solar Energy. *Trans., ASME*, **79** (6), 1349–1358 August *[60]*.

Zeuner, G. (1907). *Technical Thermodynamics*, Archibald Constable, London *[7]*. Translation by J. F. Klein of *Technische Thermodynamik*, Arthur Felix Verlag, Leipzig 1887, being edition 3 of *Grundzuege der mechanischen Waermetheorie (mit besonderer Ruecksicht of das Verfahren des Wasserdampfes)* Engelhardt (Bernhard Thierbach) Freiberg 1860.

Index

Subject Index

Index of Proper names

CPSIA information can be obtained at www.ICGtesting.com
Printed in the USA
LVOW01*1640111113

360878LV00020B/1170/P

9 780791 801710

CPSIA information can be obtained at www.ICGtesting.com
Printed in the USA
LVOW01*1640111113

360878LV00020B/1170/P

9 780791 801710